VANISHING
TREASURES

Also by Katherine Rundell

Super-Infinite: The Transformations of John Donne

Why You Should Read Children's Books,
Even Though You Are So Old and Wise

For Young Readers

Impossible Creatures

Rooftoppers

The Wolf Wilder

The Explorer

The Good Thieves

VANISHING
TREASURES

A Bestiary of
EXTRAORDINARY
ENDANGERED
CREATURES

Katherine Rundell

with Illustrations by Talya Baldwin

DOUBLEDAY  NEW YORK

Book design by Anna B. Knighton
Jacket illustration by Talya Baldwin
Jacket design by Ariel Harari

Library of Congress Cataloging-in-Publication Data
Names: Rundell, Katherine, author. | Baldwin, Talya, illustrator.
Title: Vanishing treasures : a bestiary of extraordinary endangered creatures / Katherine Rundell ; with illustrations by Talya Baldwin.
Other titles: Golden mole
Description: New York : Doubleday, [2024] | "Originally published in hardcover in Great Britain as The Golden Mole by Faber & Faber Ltd, in 2022." | Includes bibliographical references. |
Identifiers: LCCN 2023057184 (print) | LCCN 2023057185 (ebook) |
ISBN 9780385550826 (hardcover) | ISBN 9780385550833 (ebook)
Subjects: LCSH: Endangered species. | Extinct animals.
Classification: LCC QL82 .R863 2024 (print) | LCC QL82 (ebook) |
DDC 591.68—dc23/eng/20240607
LC record available at https://lccn.loc.gov/2023057184
LC ebook record available at https://lccn.loc.gov/2023057185

MANUFACTURED IN THE UNITED STATES OF AMERICA

1st Printing

To my uncle Chris,

who taught me a great deal

about the living world

The world will never starve for want of wonders;
but only for want of wonder.

G. K. CHESTERTON

Contents

VANISHING
TREASURES

Introduction

A common swift, in its lifetime, flies about 1.2 million miles; enough to fly to the moon and back twice over, and then once more to the moon. For at least ten months of every year, it never ceases flying; sky-washed, sleeping on the wing, it has no need to land.

The American wood frog gets through winter by allowing itself to freeze solid. Its heart slows, then stops altogether: the water around its organs turns to ice. Come spring, it thaws, and the heart kick-starts itself spontaneously into life. We still don't understand how the heart knows to start beating.

At sea, dolphins whistle to their young in the womb; for months before the birth, and for two weeks afterward, the mother sings the same signature whistle over and over. The other dolphins are quieter than usual for those weeks, in a bid not to confuse the unborn calf as it learns its mother's call.

. . .

These things—everlasting flight, a self-galvanizing heart, a baby who learns names in the womb—sound like fables we tell children. But it's only that the real world is so startling that our capacity for wonder, huge as it is, can barely skim the edges of the truth.

This book is made in part of moments where we have collided with living things, in joy and destruction, grandeur and folly. They are histories that reveal us to ourselves, and which find us at our most enthralled and unhinged: at our strangest. That, for instance, St. Cuthbert, a seventh-century monk from Lindisfarne, was said to have enlisted the help of sea otters when he got wet in the ocean: they warmed his feet with their breath, and dried them with their fur. That a beautiful young woman told Alexandre Dumas that she would gladly go to bed with him, but only if he would first give her as a love-offering a mongoose and an anteater. That a blind farmer in Suriname once rescued a baby capybara and trained it to be his seeing eye. It was noted in the *Guinness Book of World Records:* guided by what is essentially a vast guinea pig, a man once stepped bravely out into the darkness of the world, and was led home.

History does not relate whether Dumas was able to follow through on the deal: it seems unlikely. The mongoose would have been easy enough to buy in nineteenth-century Paris, but the anteater far less so. What's not difficult to know, though, is why the young woman wanted the elegant rat-cat and the worm-tongued mammal in the first place: we have had such hunger for the living creatures with which we share the world.

This book is, too, a litany of the many wild guesses and misunderstandings, the vivid mistakes upon which our knowledge has been painstakingly built. For example: because we used to hunt beavers for their testicles on the grounds that they were a delicious aphrodisiac, we theorized for hundreds of years that if chased, the animal would bite off its own genitals in order to forestall the pursuit. They would "throw them in [their pursuers'] path," a Roman text from the year 200 CE claimed, "as a prudent man who, falling into the hands of robbers, sacrifices all that he is carrying, to save his life, and forfeits his possessions by way of ransom." Medieval bestiaries were therefore populated with images of furious beavers castrating themselves with their incisors.

Similarly, the medieval conviction that ostriches could digest iron meant that Arab and European manuscripts were scattered with drawings of the bird with a horseshoe or a sword clamped hungrily in its beak. The theory was tested and recorded by the great ninth-century Iraqi naturalist al-Jahiz, who reported that the ostrich happily ate burning pieces of metal, but upon devouring a pair of scissors, sliced itself open from the inside. We also believed that ostriches could hatch their eggs merely by glaring at them with great and unswerving intensity.

The old errors are fantastical and fantastic, and revealing of human hopes and anxieties: our terrors, our desires for greater digestive health and sexual prowess, our quest for magical solutions to relentlessly human problems. And every scientist you meet will tell you: there is no reason to believe that we haven't got just as much wrong today as we have done in every generation up till now. It would be worth our holding that knowledge, tight and urgent, as we go; our learning, though vast, is an infinitesimally small fraction of what exists.

We risk losing all this magnificence before we begin to understand it. Every species in this book is endangered or contains a subspecies that is endangered—because there is almost no creature on the planet, now, for which that is not the case. It is the global West that has contributed most to the destruction of the world's ecosystems, but the impact will be felt by everything on Earth. Time is running short. In the last fifty years, the world's wildlife has declined by an average of almost seventy percent. We have lost more than half of all wild things that lived.

We are Noah's Ark in reverse: it is as if we are raging through the bowels of the boat, setting fire to the stables, poisoning the water. Faced with such destruction at such pace, acquiescence becomes impossible. The time to fight, with all our ingenuity and tenacity, and love and fury, is now.

THE

Wombat

T he Wombat," Dante Gabriel Rossetti wrote in 1869, "is a Joy, a Triumph, a Delight, a Madness!" The painter's house at 16 Cheyne Walk in Chelsea had a large garden, which, shortly after he was widowed, he began to stock with wild animals. He acquired, among other beasts, wallabies, kangaroos, a raccoon, and a zebu. He looked into the possibility of keeping an African elephant but concluded that at £400 it was unreasonably priced. He bought a toucan, which, it was rumored, he trained to ride a llama. But, above all, he loved wombats.

He had two, one named Top after William Morris, whose nickname "Topsy" came from his head of tight curls. In September 1869, Rossetti wrote in a letter that the wombat had successfully interrupted a seemingly uninterruptable monologue by John Ruskin by burrowing its nose between the critic's waistcoat and jacket. Rossetti drew the wombats repeatedly; he sketched his mistress—William Morris's wife, Jane—walking one on a leash. In the image, both Jane and the wombat look irate. Both wear halos.

It isn't difficult to understand Rossetti's devotion. Wombats are deceptive; they are swifter than they look, braver than they look, tougher than they look. Outwardly, they are sweet-faced and rotund. The earliest recorded description of the wombat came from a settler, John Price, in 1798, on a visit to New South Wales. Price wrote that it was "an animal about twenty inches high, with short legs and a thick body with a large head, round ears, and very small eyes; is very fat, and has much the appearance of a badger." The description implies only limited familiarity with badgers; in fact, a wombat looks somewhere between a capybara, a koala, and a bear cub. And though most are a serviceable brown, a small number of southern hairy-nosed wombats are born with a rare genetic mutation which makes their fur gold, the rich blonde of Marilyn Monroe.

Despite the fact that they do not look streamlined, a wombat can run at up to twenty-five miles an hour, and maintain that speed for ninety seconds. The fastest recorded human foot speed was Usain Bolt's hundred-meter sprint in 2009, in which he hit a speed of 27.8 miles per hour but maintained it for just 1.61 seconds, suggesting that a wombat could readily outrun him. Wombats can also fell a grown man, and have the capacity to attack backward, crushing predators against the walls of their dens with the bone-hard cartilage of their rumps. The shattered skulls of foxes have been found in wombat burrows.

Wombats are careful and protective mothers, giving birth once a year in the spring. Like all marsupials, they produce tiny embryonic young, born after only thirty days in utero, which then take refuge in the mother's pouch for eight months and develop further. The wombat's pouch is positioned upside down, so the joey's head looks out between the mother's hind legs, in order to allow her to dig without filling the pouch with mud. It's an extraordinary bit of animal adaptation that also makes it seem as though the mother wombat is in a state of constant, eight-month labor, which explains why it was a kangaroo who got to be in *Winnie the Pooh*.

To early settlers in Australia, wombats were pests. Although wombat hams, like kangaroo steaks, could supplement the scant settler diet, they were primarily regarded as a potential threat to crops and were slaughtered en masse. (Their spoor is easy to track, for their droppings take the form of almost perfect cubes.) In Victoria in 1906, wombats were classed as vermin; in 1925, a bounty was introduced, and hunters could make ten shillings per wombat scalp. The bounty incentivized hunting; in one year, more than a thousand wombat scalps were traded in by a single landowner. Now, despite its name, the common wombat is no longer common. Overgrazing and the destruction of their natural habitat has caused a sharp drop in their numbers; all species of wombat are now protected, and the northern hairy-nosed wombat is critically endangered. It has sleeker, softer fur than the common variety, and poor eyesight, relying on its large silky nose to guide it to food in the dark. As its habitat has slowly been cut away, it has become one of the rarest land mammals in the world. A census in 1982 put the surviving number at 30; the most recent found that 251 northern hairy-nosed wombats had evaded the clumsy destructions of humankind.

Other wombats have died more directly at human hands. In 1803, Nicolas Baudin, the famed explorer, returned from a journey to New Holland (now Australia) with an ark of animals for Napoleon's wife, Josephine. The voyage was bleak, with a high death count among its company and cargo: more than half the crew had to abandon ship owing to illness, ten kangaroos died of exposure, and the resident botanist had his room dismantled to make an indoor space for the remaining animals. The sick emus were fed sugar and wine and grew sicker, and Baudin himself began spitting blood. Two wombats died, but at least one other was delivered into the arms of the Empress Josephine.

Wombats have offered solace where little other solace could be found. The German philosopher Theodor Adorno was a frequent visitor to Frankfurt Zoo after the Second World War. He wrote to the director in 1965: "Would it not be nice"—or *beautiful: "wäre es nicht schön"*—"if Frankfurt Zoo could acquire a pair of wombats? . . . From my childhood I remember great feelings of identification with these friendly rotund animals, and would be filled with delight to see them again."

It is not always enough to be loved. Rossetti's wombats did not thrive in captivity. His last wombat sketch is of himself, his handkerchief covering his face, weeping over the dead body of a wombat. Below it, he wrote a mournful quatrain:

I never reared a young Wombat
To glad me with his pin-hole eye
But when he most was sweet and fat
And tail-less; he was sure to die!

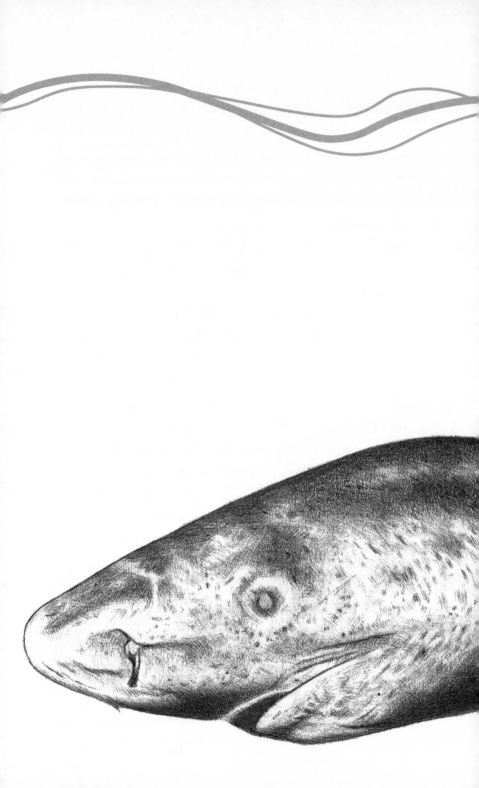

THE

Greenland
Shark

I n 1606 a devastating pestilence swept through London; the dying were boarded up in their homes with their families, and a decree went out that the theaters, the bear-baiting, and the brothels be closed. It was then that Shakespeare wrote one of his very few references to the plague, catching at our precarity:

> *The dead man's knell*
> *Is there scarce asked for who, and good men's lives*
> *Expire before the flowers in their caps*
> *Dying or ere they sicken.*

As he wrote the words, a Greenland shark who is still alive today swam untroubled through the waters of the northern seas. It was, at the time, perhaps a hundred years old, still some way off its sexual maturity: its parents would have been old enough to have lived alongside Boccaccio: its great-great-grandparents alongside Julius Caesar. For thousands of years Greenland sharks have swum in silence, as aboveground the world has burned, rebuilt, burned again.

The Greenland shark is the planet's oldest vertebrate, but it was only recently that scientists were able to ascertain exactly how old. A Danish physicist, Jan Heinemeier, discovered a way to test lens crystallines, proteins found in the eye, for carbon-14. The amount of carbon-14, a radioactive isotope, found naturally on Earth varies from year to year; there were huge spikes during the 1960s, when mankind was at its most enthusiastic about nuclear weapons, but every period has its own carbon-14 signature. By testing the crystallines in the sharks' eyes, it was possible to determine, very roughly, their date of birth: of twenty-eight tested, the largest, a 16.4-foot female, was reckoned to be somewhere between 272 and 512 years old. Size is thought a relatively good indicator of age, and there are records of Greenland sharks reaching 23 feet long; so it's very possible that in the water today there are sharks who are well into their sixth century.

The Greenland shark is not obviously beautiful. Its face is blunt, its fins stunted, and its eyes attract a long worm-like crustacean, *Ommatokoita elongata*. These attach themselves to the shark's corneas, fluttering from its eyeballs like paper streamers, rendering it both almost blind and more disgusting than seems fair. It smells, too. Its body has high concentrations of urea, a necessity to ensure its body has the same salt concentration as the ocean, preventing it from losing or gaining water through osmosis, but it is a necessity that means it smells of pee—so much so that in Inuit legend, the shark is said to have arisen from the urine pot of Sedna, goddess of the sea. The urea is also what makes it poisonous to humans when eaten fresh. If raw and untreated, the toxins in the flesh can render you "shark drunk": giddy, staggering, slurring, vomiting. It becomes safe only if the meat is buried for several months and left to ferment, then hung out to dry for months more. Served in small chunks, and known as *hákarl*, it is considered, by some, a delicacy, and by others an abomination. Apparently it tastes like a very ripe cheese, left for a week in high summer in a teen-age boy's car.

The Greenland shark is slow, as befits a fish so venerable. At full speed and with strenuous effort, it moves somewhere between 1.7 and 2.2 miles per hour. Although one of the two largest flesh-eating creatures in the sea, it has an astonishingly slow metabolism; in order to survive, a 440-pound shark would have to consume the calorific equivalent of one and a half chocolate cookies per day. They are hungrier in the womb than in their waking lives: the strongest fetus develops sharp teeth, eats its siblings, and emerges into the water alone. Once born, they're both hunters and scavengers; they have been thought to hunt seals, perhaps inhaling them as they sleep on the surface of the water, but largely they eat whatever falls off the ice: reindeer, polar bears. The leg of a man was found in one shark's stomach, although none of the rest of him. And the Greenland shark is slow even in the process of its dying. Henry Dewhurst, a ship's surgeon writing in 1834, saw a shark caught and killed:

When hoisted upon deck, it beats so violently with its tail, that it is dangerous to be near it, and the seamen generally dispatch it, without much loss of time. The pieces that are cut off exhibit a contraction of their muscular fibres for some time after life is extinct. It is, therefore, extremely difficult to kill, and unsafe to trust the hand within its mouth, even when the head is cut off . . . This motion is to be observed three days after, if the part is trod on or struck.

They live deep-down and secret lives. Although they have been seen at the water's surface, they prefer to be close to the bottom of the ocean, where it's dark and cold: they've been found as far down as 7,200 feet, more than seven Eiffel Towers deep. Nobody has ever seen one give birth; we have never seen them mate. Their invisibility to us also means that we do not know how endangered they are: they're currently listed as "near threatened," but they could be the most common sharks in the world, or urgently at risk. We do know that for some time they were overfished in large numbers—thirty thousand a year in the 1900s—in order to extract oil from their bodies. It was said that there were places in the Norwegian archipelago where houses decorated in the paint made from the shark's liver oil fifty years ago still shone bright: a paint like no other. We know, too, that because it takes 150 years for a female to be ready to breed, they replenish slowly. They were also believed to be excellent parents: the second-century Greek poet Oppian averred that, when threatened with danger, a parent shark would open her cavernous mouth and conceal her young ones within. As this is very much, alas, not likely to be true, we will need to take care of them ourselves.

Because they live so far beyond our ships and divers, we do not know where they swim. They come to the surface only in the places where it is cold enough for them, in the Arctic, around Greenland and Iceland, but they have been found in the depths near France, Portugal, Scotland. Scientists say they may be everywhere the ocean goes deep enough and cold enough: they could be far closer to us than we think.

I am glad not to be a Greenland shark; I don't have enough thoughts to fill five hundred years. But I find the very idea of them hopeful. They will see us pass through whichever spinning chaos we may currently be living through, and the crash that will come after it, and they will live through the currently unimagined things that will come after that: the transformations, the revelations, the possible liberations. That is their beauty, and it's breathtaking: they go on. These slow, odorous, half-blind creatures are perhaps the closest thing to eternal this planet has to offer.

THE

Raccoon

Rebecca was an unusual White House inhabitant for two reasons: The first was that she arrived as a prospective dinner. The second, and most immediately obvious on meeting her, was that she was a raccoon.

In 1926, a citizen of Mississippi sent the raccoon to the First Family in time to be cooked for Thanksgiving. Calvin Coolidge, who (the White House was at pains to make clear) had never eaten raccoon, instead kept her as a pet, adding her to his famous menagerie of animals, which included a pygmy hippopotamus and a bear. The First Lady, Grace Coolidge, wrote of Rebecca: "We had a house made for her in one of the large trees, with a wire fence built around it for protection." But Rebecca spent most of her time inside, where "she had her liberty. She was a mischievous, inquisitive party and we had to keep watch of her when she was in the house. She enjoyed nothing better than being placed in a bathtub with a little water in it and given a cake of soap with which to play. In this fashion she would amuse herself for an hour or more."

She was, as befitted the First Raccoon, exquisitely accessorized: she wore an embroidered collar with an engraved nameplate. Clad in her finery, she roamed the White House, unscrewing light bulbs and upending houseplants from their pots. She participated in the Easter Egg Roll on the White House lawn, a satin bow tied to her collar. Coolidge, the press reported, liked to have her in his study, sometimes draped around his neck, stroking her as he worked late into the night. And so she lived a life of luxury, until she did a thing many of her fellow Americans have dreamed of and very few achieved: she bit the president of the United States. At least, historians assume so: Coolidge was seen in public with a bandaged hand, and Rebecca was temporarily sent to the zoo. *The Baltimore Sun* reported: "From White House to Zoo Is Rebecca's Sad Story."

She was allowed, after a period in exile, to return, and efforts were made to provide her with a companion. In 1928, a White House police officer captured a wild male raccoon and offered him as a playmate: the raccoon, named Reuben, ran away, and Rebecca lived on in stately solitude. Grace Coolidge wrote: "Rebecca had lived alone and had her own way so long that I fear she was a little overbearing and dictatorial, perhaps reminding her spouse that he was living on her bounty."

Outside the White House, where there are no staff to provide soap and snacks, the raccoon is a formidable survivalist. While their numbers in America in the 1930s looked on the edge of collapse, they learned to adapt to our urban environments, and came roaring back: now the common raccoon is thought to be one of the most populous mammals in North America, to which it is native. Raccoons are rare in being drawn to unfamiliar landscapes rather than familiar ones: they are neophiles, explorers. Their faces, banded with markings— which suggest, depending on the subspecies, anything from "burglarious intent" to "masked black-tie ball"—appear vividly inquisitive, and their instincts match. And even people who are hounded by raccoons—who find their garbage cans upended and small children terrified—admit their astonishing intelligence. Trash bandits they may be, but they are not to blame for the infelicities they cause us: one must salute both their beauty and their flair.

Their paws are hypersensitive, having a form of "whisker" (the vibrissae) on the front claws and a palm which becomes supple when wet; where possible, a raccoon will dunk its food underwater, the better to study it. This spinning of food in its forepaws led to its name, a derivation from the Powhatan word *arakun*, meaning "creature that scratches with its hands." Their remarkable manual dexterity allows them to take lids off trash cans, fish food from tins, lift latches, and push open windows. Their creative reasoning is remarkable: a study at the University of Wyoming found that they excelled at the "Aesop test," in which animals are presented with a cylinder containing a marshmallow floating on water that is too low to reach. A human demonstrated how dropping pebbles into the water raised the water level. Two of the eight raccoons successfully copied the human behavior almost immediately: another jumped on the cylinder and tipped it over, so that the marshmallow could be snatched—something no other test subjects had thought of. In a famous early experiment in 1913, Walter Hunter tested the memory of a number of mammals, and found that in many circumstances raccoons performed better than dogs. An Instagram-famous raccoon in the UK called Melanie has been trained by her American owner to pedal a bicycle and sweep the floor with a broom: she competed on *Britain's Got Talent*. From time to time she wears a hat, a tutu, or a bowtie, about which she presumably has mixed feelings, but which television executives enjoy unabatedly.

But for all their beauty and intelligence, the raccoon has also been connected to some of America's bleakest and darkest history; *coon* was, and remains, a vicious racist slur. Slurs are rarely simple, and the route by which *coon* became a violently degrading anti-Black epithet is a tangled one. It was initially, in the mid-eighteenth century, associated with whites, referring to the shrewd, raccoon-like survival of rural woodsmen, used to salute the independent verve of the frontier. It had such a tinge of roguish rebellion to it that Davy Crockett was referred to as a "right smart coon" in 1832. But in 1834, the songwriter George Washington Dixon popularized the character of Zip Coon: a caricature of free Black Americans, imagined by a white man in burnt-cork blackface as arrogant, urban, and aggressive. Zip Coon—dandified and ostentatious, yet portrayed as ignorant and uncivilized—spread like a plague through white American media and song: he became a perfect channel for racist fury and terror at Black American advancement in the run-up to the Civil War.

Meanwhile, raccoons themselves became less beloved: portrayals of them in the news became less centered on their roguish charm and more about their thieving. Added to that, white lore about Black people having a fondness for eating raccoons grew (it had roots in truth: for the enslaved, raccoon was a source of sustenance outside of the deliberately constricted rations they were provided by their enslavers). By the end of the nineteenth century, the association of Black Americans with *coon* had become widespread and vicious: one piece of a long, dark history that is still being reckoned with, and for which justice still remains to be served.

Left to themselves, raccoons are a large and vibrant band: there are three species and at least twenty-two subspecies. And though the common raccoon is very common indeed, the Cozumel raccoon, the smallest of the species, is one of the most endangered animals in the world. Also known as the "pygmy" or "dwarf" raccoon, perhaps three hundred are left in the wild. Found only on the island of Cozumel off the coast of Mexico, they hunt for crabs and other sea life in the shallows, spinning lobsters and crayfish in their paws. Our encroachments into the mangrove forests in the northwest of the island, driven by the tourism industry, have made their survival more peril- ous every year: unless laws are passed to protect the forests, the animals will soon be gone. Just half the size of the com- mon raccoon, with a rich russet tinge to the tail, they move, sometimes alone and sometimes in groups, along the water's edge, searching for food. With their sharply tuned faces and smart gray uniforms, they look like a team of diminutive pri- vate investigators.

If you were looking for charm, you might turn to the Torch Key raccoon, found in Florida—one of the smallest and light- est raccoons, the ochre-gray color of Parisian limestone—or to the Upper Mississippi Valley raccoon, one of the largest and darkest of the subspecies, which can be found as far north as

Wisconsin and has unusually luxuriant fur to help it through the snow-thick winters. And then there is the Barbados raccoon: one of the very fairest, it is petite and delicate. A subspecies of the common raccoon, its skull is delicately slender, and its fur shades to a soft yellow-white about the face.

But there are no more Barbados raccoons. Their numbers were whittled away by traffic and by human incursion, driven by tourism, into the spaces they relied upon for their existence. The last known member of the family was hit by a car in 1964. In 1996, they were officially declared extinct. They were very fine, and now we have only a few taxidermied skins left.

· · ·

It is a difficult world to survive in, this one we have created. It is so very beautiful; and it is so ruthlessly perilous to the vulnerable. We will have to demand more, both from our adversaries and from our friends. The alternative—a world in which humanity lives alongside a tiny handful of species, amid empty skies and empty seas—becomes more possible every year. Now is the time to act: now, and now, and now.

THE

Giraffe

The Roman poet Horace was stridently anti-giraffe. The animal was, he believed, conceptually untidy: it was, he said, a "hybrid monster . . . half camel, half leopard." He wrote in his account of nature in *Ars Poetica* (c. 8 BCE): "If a painter had chosen to set a human head on a horse's neck [or] if a lovely woman ended repulsively in the tail of a black fish, could you stifle laughter, friends?" When Julius Caesar brought a giraffe back to Rome from Alexandria in 46 BCE (a gift, some said, from Cleopatra), those lining the streets saw, as Horace did, a creature made of two parts. Cassius Dio wrote in his *Historia Romana* that it was "like a camel in all respects except that its legs are not all of the same length, the hind legs being the shorter . . . Towering high aloft, it . . . lifts its neck in turn to an unusual height. Its skin is spotted like a leopard." But the crowds rejoiced in the creature's bravura hybridity. "And for this reason," Dio wrote, "it bears the joint name of both animals." *Camelopardalis:* camelopard.

Throughout history we have tried, with more enthusiasm than accuracy, to explain how something so mixed and miraculous came to be. The Persian geographer Ibn al-Faqih wrote in 903 that the giraffe occurs when "the panther mates with the [camel mare]." Zakariya al-Qazwini, a cosmographer from the thirteenth century, suggested in his *Wonders of Creation* (which also includes among its marvels *al-mi'raj*, a rabbit with the horn of a unicorn) that its genesis was the result of a two-part concatenation: "The male hyena mates with the female Abyssinian camel; if the young one is a male and covers the wild cow, it will produce a giraffe." Both possibilities sound more stressful than would be ideal from an evolutionary point of view. Others have declared it magical: the early Ming dynasty explorer Zheng He brought two giraffes to Nanjing and cherished them as *qilin*, a gentle hooved chimera. Charles I's chaplain, Alexander Ross, wrote in *Arcana Microcosmi* in 1651 that the sheer fact of the giraffe made it impossible for naturalists to "overthrow the received opinion of the ancients concerning griffins . . . seeing there is a possibility in nature for such a compounded animal. For the gyraffa, or camelopardalis, is of a stranger composition, being made of the leopard, buffalo, hart and camel."

Ross was right: the truths of the giraffe are more fabulous and potent than our fictions. Giraffes are born with no aid from the camel or hyena, but even so their birth is a wonder: they gestate for fifteen months, then drop into existence a distance of five feet from the womb to the earth. It looks as brisk and simple as emptying out a handbag. Within minutes, they can stand on their trembling, catwalk-model legs and suckle at their mother's four teats, biting off the little wax caps that have formed in the preceding days to keep the milk from leaking out. Soon they are ready to run, but still liable to trip over their own hind legs, a hazard they never learn to avoid entirely.

Once full grown, they can gallop at thirty-five miles an hour on feet the size of dinner plates, but it remains safer not to: they self-entangle. Their tongue, which is dark purplish-blue to protect it from the sun and more powerful than that of any other ungulate, is almost twenty inches long: they can scrape the mucus from deep inside their own nostrils with the tip. And they are the skyscrapers of mammals, unmatched: the tallest giraffe ever recorded, a Masai bull, measured 19 feet tall. The explorer John Mandeville only mildly exaggerated when he wrote of the "gerfauntz," in the first English-language account in 1356, that it had a neck "twenty cubytes long [about thirty feet] . . . he may loken over a gret high hous." (As Mandeville is itself a fictional appellation for an unknown man, some laxity in measurements is to be expected.) But though so tall, they are hospitable to the small. They have been known to host tiny yellow-billed oxpeckers on their bodies: the small birds remove ticks from their skin, and clean the food from between their teeth. Giraffes have been photographed at night with clusters of sleeping birds tucked into their armpits, keeping them dry.

In Atlanta, Georgia, it is illegal to tie your giraffe to a street-lamp. It is not illegal, though, to import a cushion made from a freshly shot giraffe's head with the eyelashes still attached. The United States is one of the largest markets in the world for giraffe parts, because America has refused to designate the animals endangered, despite the fact that there are fewer than 68,000 left in the wild, a forty percent drop in thirty years. In a recent ten-year period, American hunters imported 3,744 dead giraffes—about five percent of the total number alive. Today you could, if you felt like externalizing the apocalyptic whiff of your personality, buy both a floor-length giraffe coat and a Bible in a giraffe-skin cover. Rarer breeds are on the very edge of vanishing: the population of Nubian giraffe has fallen by ninety-eight percent in the last four decades, and they will soon be extinct in the wild. Their own beauty imperils them. As the great Roman naturalist Pliny wrote, the proof of wealth is "to possess something that might be absolutely destroyed in a moment."

We don't know why the giraffe looks as it does. Until relatively recently, its neck was explained in the way Darwin suggested: the "competing browsers hypothesis" posits, commonsensically, that competition from browsers such as impala and kudu encouraged the gradual lengthening of the neck, allowing it to reach food the others couldn't. Recently, though, it has been shown that giraffes spend relatively little time browsing at full height, and the longer-necked individuals are more likely to die in times of famine. It is possible that it gives males an advantage when they engage in "necking"—swinging their necks against one another, seemingly to establish dominance. (There will surely be more to discover about necking, too, in years to come: it often leads to sexual activity between the warring males. Indeed, most sex between giraffes is homosexual: in one study, same-sex male mounting accounted for ninety-four percent of all sexual behavior observed.) Whatever its reason, the neck comes at a price. Each time a giraffe dips down to drink, its legs splayed, blood rushes to its brain; as it bends, the jugular vein closes off blood to the head to stop it from fainting when it straightens up again. Even when water is plentiful, they drink only every few days. It is a dizzying thing, being a giraffe.

There is something in giraffes that unhinges us in our delight. In 1827, a giraffe walked into Paris. She was not the first giraffe in Europe—Lorenzo de' Medici had brought a giraffe through Italy in 1487, Florentines leaning perilously out of second-floor windows to feed it—but she was the best dressed. Wearing a two-piece custom-made raincoat embroidered with fleurs-de-lis, she was a gift from the Egyptian ruler Muhammad Ali to Charles X. She traveled for more than two years from Sennar by boat and on foot, arriving in Paris in high summer; there she bent to eat rose petals from the king's hand. She was known as *la Belle Africaine, le bel animal du roi*, and, most often, *la girafe:* like God and the king, there was only one. She was housed in the royal menagerie in an enclosure with a polished parquet floor ("truly the boudoir of a little lady," the keeper wrote) and Parisians, filing past to see her in their thousands, went giraffe crazy. Shops filled with giraffe porcelain, soap, wallpaper, cravats, giraffe-print dresses; the colors of the year were "Giraffe belly," "Giraffe in love," and "Giraffe in exile." Hair was worn vertically in Paris that season. Women smeared their hair with hogs' lard pomade fragranced with orange flower and jasmine, and wound it to resemble the giraffe's "horns," or ossicones. There were reports of women sitting on the floors of their carriages, so high did their *coiffures à la girafe* rise.

But we tire of everything, even miracles. Charles X abdicated, his son ruled for twenty minutes, and *la girafe* outlived her fame. She died unvisited in 1845, was taxidermied and put in the foyer of the Jardin des Plantes. Delacroix went to see the body: the giraffe, he wrote, died "in obscurity as complete as her entry in the world had been brilliant." But the wild Parisian reaction, it seems to me, was the only reasonable one. It should never have died down: we should still be wearing our hair in twelve-inch towers. Why did we ever stop? The earth is so glorious and so unlikely: the giraffe, stranger than the griffin, taller than a great high house, offers us the incomparable gift of being proof of it.

THE

Swift

The swift is sky-suited like no other bird. Weighing less than a hen's egg, with wings like a scythe and a tail like a fork, it eats and sleeps on the wing. They gather nesting material only from what's in the air, which means that there have been accounts of still-flapping butterflies wedged in among the leaves and twigs. They mate in brief mid-sky collisions, the only birds to do so, and to wash they hunt down clouds and fly through gentle rain, slowly, wings outstretched.

Most remarkable of all is their night. Swifts can find a state of unihemispheric sleep; they shut off one half of their brain at a time, while the other remains functioning, alert to changes in the wind, so that the bird wakes in exactly the same place where it fell asleep; or, if migrating, on the precise course it set itself. The left side closes first, then the right, so that it sways a little in the air as it sleeps. Geoffrey Chaucer knew it long before we did: in *The Canterbury Tales* he wrote about small birds who "slepen al the nyght with open ye." And a French pilot during the First World War, flying by the light of a full moon on a reconnaissance mission near the Vosges, saw a ghostly cloud of them, apparently hovering entirely still in the air:

> As we came to about ten thousand feet . . . we suddenly found ourselves among a strange flight of birds which seemed to be motionless, or at least showed no noticeable reaction. They were widely scattered and only a few yards below the aircraft, show-

ing up against a white sea of cloud underneath. None was visible above us. We were soon in the middle of the flock.

Nobody believed him at the time: it seemed impossible, because swifts do seem impossible.

The swift is of the family Apodidae, from the Greek *ápous,* "footless," because they were once believed to have no legs. We still know very little about them, because they're so hard to catch and study, but we do at least know that they have legs, albeit tiny, weak ones. Adult swifts can walk if they absolutely have to, but younglings can't, and, if all goes well, they do not need to. They tip themselves from the nest and fly straight to Africa, some not alighting again for ten months, some for two years or four, and a few never stopping at all. We know that to prepare for their great flight, the young chicks in the nest strengthen their wings from a month old by doing feathery push-ups; lifting their bodies up off the nest by pressing down on their wings, until they can hold themselves there, suspended, for several seconds. Then they're ready.

We still don't know with absolute certainty how they know with such absolute certainty where to go. But we know that they're fast. They are the swiftest of all birds in level flight (a peregrine can outstrip them in a dive, but they can outfly her in a flat race); the top speed ever officially recorded was almost 70 miles per hour, but there are reports of the needle-tailed swift, found in Africa and Asia, reaching 105 miles per hour. A swift flies about 124,000 miles a year; the Earth has a circumference at the equator of 24,901 miles; so a swift flies far enough each year to put five girdles around the Earth. It would be exhausting to contemplate, except that I have never seen a swift that looks exhausted.

That sky-high stamina—and their raucous, questing cry—has long thrilled us. In heraldry, the swift is one of the inspirations for the imaginary martlet, a stylized bird without feet. Unable to land, the martlet is a symbol of restlessness and pursuit: of the constant search for knowledge and adventure and learning. They were used in coats of arms as the mark of the fourth son, on the reckoning that the first son got the money, the second and third went to the Church, and the fourth was free to seek his fortune. Edward the Confessor, a king so morally upright he was practically levitating, was given five martlets, posthumously, on his shield: gold, to match the sun they fly after.

The Apodidae family is ancient: it separated off from other birds about seventy million years ago, so they're old enough as a species to have had a nodding acquaintance with the *Tyrannosaurus*. They evolved to have deep-set eyes with bristles in front, which act as sunglasses against the glare of the equatorial sun. There are more than a hundred species, from the tiny pygmy swiftlet, just three and a half inches across and found only in the Philippines, to the white-naped swift, a huge ten inches across the wing, which is silent when alone and calls almost ceaselessly when with its flock, *cree cree cree*.

If you see a bird settled on a telegraph wire or a tree, it's not a swift. No sociable windowsill singers, no Disney Princess finger-perchers, they fly wild, and they fly like a stroke of luck incarnate. But they would, like most living things, be far luckier without us; the only swift found in Britain, *Apus apus*, which arrives in the UK for the mating season, is not yet critically endangered, but the last two decades have seen a fifty percent drop in their breeding numbers. *Apus apus* mate for life, returning every year from Africa to the same spot to nest, usually in spaces under roof tiles and in the eaves of old houses and barns. As we knock down and seal up old buildings, the swifts can find nowhere safe to lay their eggs before the breeding season is over. North America's most common swift, the lovely soot-gray chimney swift, *chaetura pelagica*, known as "the cigar with wings," has also been plummeting in numbers: they are now listed as "vulnerable." They, like swifts the world over, suffer from mass industrialized pesticide use and global warming, both of which affect the insect population; swifts can eat only what's in the air, and a swift with chicks needs to gather as many as a hundred thousand insects a day, storing them in batches of a thousand in a bulge in their throat. The world's ever-more-frequent extreme weather events will decimate populations: after 2005's Hurricane Wilma, the Canadian swift population was halved almost overnight. And then there's our own familiar deadly hunger, our compulsion to feast at another's expense: bird's nest soup, thought to clear the complexion and rejuvenate the body, requires the harvesting of vast numbers of the endangered swiftlets' nests. These days, the swift's cry could quite reasonably be heard as a warning, or a sharp and angry accusation.

Ted Hughes wrote a love poem to swifts in the 1970s. It catches at their glory, their racing high-pitched valiance, although it reads a little differently now:

They've made it again,
Which means the globe's still working, the Creation's
Still waking refreshed, our summer's
Still all to come—

THE

Lemur

I t is probably best not to take advice direct and unfiltered from the animal kingdom—but lemurs are, I think, an exception. They live in matriarchal troops, with an alpha female at their head. When ring-tailed lemurs are cold or frightened, or when they want to bond, they group together in a furry mass known as a lemur ball, forming a black-and-white sphere that ranges in size from a football to a bicycle wheel. They intertwine their tails and paws, and press against one another's walnut-sized swiftly beating hearts. To see it feels like an injunction of sorts: to find a lemur ball of one's own.

The first lemur I ever met was a female, and she tried to bite me, which was fair, because I was trying to touch her, and because humans have done nothing to recommend themselves to lemurs. She was an indri lemur, living in a wildlife sanctuary outside the capital of Madagascar, Antananarivo; she had an infant, which was riding not on her front, like a baby monkey, but on her back, like a miniature Bill Shoemaker. She had wide yellow eyes. William Burroughs, in his lemur-centric eco-surrealist novella *Ghost of Chance*, described the eyes of a lemur as "changing color with shifts of the light: obsidian, emerald, ruby, opal, amethyst, diamond." The stare of this indri resembled that of a chemically enhanced young man at a nightclub who urgently wishes to tell you about his belief system, but her fur was the softest thing I have ever touched. I was a child, and the indri, which is the largest extant species of lemur, came up to my ribs when standing on her hind legs.

She looked, as lemurs do, like a cross between a monkey, a cat, a rat, and a human.

Lemurs are strange in the way that the reclusive and the wealthy are strange; having had the island of Madagascar to themselves to evolve in, they have idiosyncratic habits. Male ring-tailed lemurs have scent glands on their wrists, and engage in "stink-fighting," battles in which they stand two feet apart and wipe their hands on their tails, then shake the tail at their opponent, all the while maintaining an aggressive stare until one or the other retreats. It feels no madder than most current forms of diplomacy. It's not unusual for female ring-tailed lemurs to slap males across the face when they become aggressive.

There are at least 101 species and subspecies of lemur in Madagascar; there were once lemurs the size of small men, but after humans arrived on the island two thousand years ago the larger lemurs were hunted to extinction. At the smallest end of the scale is the Madame Berthe's mouse lemur, the smallest primate in the world, which weighs one ounce on average and at full stretch couldn't cover your hand. Somewhere in the middle is the northern giant mouse lemur, whose testes make up 5.5 percent of its body mass; the equivalent proportions in a man would be testicles the size of grapefruits. They are strange, then, and beautiful, and occasionally disconcerting when seen from below.

The indri lemur was right to try to bite me; more right than she knew. The early human arrivals on the island eradicated at least fifteen species of lemur. Now, largely due to deforestation, twenty-four species are critically endangered, forty-nine are endangered, and ninety-four percent of all species are threatened. Until recently there was a strong taboo on hunting lemurs. Rural traditions held eating lemur flesh to be second in horror only to human flesh; some stories told that the lemurs were human ancestors who had become lost in the Madagascan rainforest and changed themselves into lemurs to survive. Other stories told of a man who, falling to certain death from a high tree, was caught by an indri lemur and set upright on the ground. The taboo was first eroded through poverty and desperation: in rural households where lemur was eaten, the children were almost invariably found to be malnourished. Conservation can never thrive unless it works with those people who live on the land and who know it most deeply. The lemur cannot be protected until we seek out ways to assure health and dignity for those who dwell alongside them; a by-product will be the eradication of the need to hunt endangered creatures for survival.

So the myths have not saved the lemurs. And when we endow anything or anyone with mystic powers, we usually end by killing it. The aye-aye lemur is thought in some areas to be able to prophesy death. They have vast eyes; large, sensitive ears; and a middle finger that's twice as long as their other digits. When the aye-aye points its middle finger at a person, they are taken to be cursed. Another story tells that it uses the long finger to puncture human hearts. As a result they're unbeloved, and hunted so relentlessly they were thought to be extinct until they were rediscovered in 1961. The word *lemur* comes from the Latin *lemures*, meaning "ghosts." It is possible that several subspecies may become exactly that: stories, preserved a hundred years from now only in photographs and stuffed specimens gathering dust.

Perhaps the most astonishing fact about lemurs is that they survived at all. Madagascar was part of Gondwanaland until 180 million years ago, when the supercontinent began to split and the island began to drift eastward from Africa. But the first lemur-like fossils date from between 62 and 65 million years ago, and appear in mainland Africa. How, then, did the lemurs reach Madagascar? There are many theories—island-hopping and land bridges among them—but the dominant theory is that the lemurs drifted there on floating rafts of vegetation. The island, too, kept drifting, so when monkeys evolved enough to eradicate lemurs on the mainland with their superior adaptiveness and aggression, somewhere between 17 and 23 million years ago, Madagascar was safely out of reach. I have seen many things that I've loved, but I don't think I'll live to see anything as fine as a raft of lemurs, sailing across the sea toward what looked, until the arrival of humans, like safety.

THE

Hermit Crab

It was, perhaps, a hermit crab that ate Amelia Earhart. For five nights after Earhart disappeared from the sky in 1937, the U.S. Navy picked up distress signals from Nikumaroro, an uninhabited island in the Western Pacific. When a rescue team reached the island a week later—it took time, since planes had to be loaded onto ships—it was deserted. But researchers on the island have since discovered human bones matching Earhart's size. Another, later team discovered the shattered glass of a woman's compact mirror, a few flakes of rouge, and a pot of anti-freckle cream—Earhart was known to hate her freckles. The bones were sent to be tested, but were lost on the way, and unless they are found we won't ever be sure whether they belonged to the valiant, hell-for-leather aviatrix with the face of a lion. But only thirteen bones were found, and the human body has 206: If they were Earhart's, where were the other 193?

Crunched, perhaps, to fragments. Nikumaroro is home to a colony of coconut hermit crabs: the world's largest land crab, so called because of its ability to crack open a coconut, maneuvering a claw into one of the nut's three eyeholes and prying it open. The oldest live to more than a hundred and grow to be up to 40 inches across: too large to fit in a bathtub, exactly the right size for a nightmare. In 2007, researchers decided to test the Earhart theory. The carcass of a small pig was offered to the crabs on the island, to see what they might have done to Earhart's dead or dying body. Following their remarkable sense of smell, they found the pig and tore it apart, making off with its bones to their burrows under the roots of the trees. Their strength is monumental; their claw grip can produce up to 3,300 newtons of force (the bite force of a wolf is around 2,200 newtons). Darwin called them "monstrous": he meant it as a compliment.

Even monsters, though, start small. Some hermit crabs inhabit the land, and others the sea, but they all begin microscopic and underwater. They're released as eggs into the ocean and hatch as unprepossessing larvae (though what larvae are prepossessing?), and it's only after several months that they are large enough to inhabit the smallest empty shell they can find. As they grow, they graduate from one scavenged shell to another, most frequently the delicately whorled shell of a sea snail, grasping its columella with claspers at the tip of its abdomen. They shed their exoskeletons, releasing into the sea a semi-transparent floating crab—a ghost. The coconut crab eventually outgrows all other shells, and begins to live uncovered on the land, but the majority of the 1,100-odd species of hermit crabs live in borrowed homes all their lives.

Hermit crabs are not, in fact, hermitical: they're sociable, often climbing on top of one another to sleep in great piles, and their group behavior is so intricately ordered that they make the politics of Renaissance courts look simplistic. When a crab comes across a new shell, it will climb into it and try it on for size. If the shell is of good quality but too big, it waits nearby for another crab to come and inspect it. If that crab also finds it too large, it joins the first crab, holding on to its claw until a queue develops—it can stretch to twenty crabs, arranged in order of size from smallest to largest, each holding on to the next: a hermit crab chorus line. When at last a crab arrives who can fit the vacant shell, the next crab in line claims the new crab's former shell, and there is a flurry of crabs climbing into their neighbor's home. The crab's abdomen is soft and vulnerable to attack while exposed, so the whole process takes place with astonishing rapidity.

They're not only foragers for homes: some are renovators. The anemone hermit crab is so called because it lifts anemones from the seabed and sticks them to its shell, where their stinging tentacles offer protection and disguise from predatory octopuses. The anemone, in symbiotic turn, consumes scraps of the hermit's food as they float by. When the time comes to move to a larger shell, the crab, with some difficulty and great persistence, pries its anemones off the old shell and fixes them to the new.

Because they are small, and because their eyes-on-stalks, their ommatophores, are curious and gentle, the Caribbean and Ecuadorian hermit crabs are often sold as easy-care pets. Salesmen paint their shells in bright colors, which slowly poisons them. Many, needing dense humidity to breathe, suffocate in their tanks. On the beaches, they are trapped and killed by plastic. Nor are they safe in the sea: some live at depths of more than sixty-five hundred feet, and our pollution reaches them even there. The coconut crab risks extinction in large part because its flesh is believed to be an aphrodisiac. In this faith, as with tiger claws and rhino horn, there is evidence of great human vulnerability, and enough stupidity to destroy entire ecosystems.

In fact, the sum total of natural aphrodisiacs—non-medical sexual stimulants—is zero. Historically, we've chosen to believe there are aphrodisiacal powers in (a) that which is rare, exotic, new, or expensive, or (b) food laden with spice, which speeds the metabolism and sparks heat within the body, or (c) food which looks like a penis or a vagina, or (d) food which actually *is* a penis or vagina, or eggs or similar. Oysters, for instance, are made up largely of water, protein, salt, zinc, iron, and tiny amounts of calcium and potassium: they're no more an aphrodisiac than a vitamin pill drowned in saltwater, but they look suggestive. We have in the past given sexual potency, haphazardly, to chocolate, asparagus, carrots, honey, nettles, mustard, and sparrows. In the thirteenth century the German saint Albertus Magnus posited that you might grind up badger flesh and sprinkle it on food for instant eroticism. For Shakespeare, the potato was rare and exotic, and generally believed to have aphrodisiac qualities: "Let the sky rain potatoes," Falstaff says in *The Merry Wives of Windsor*, "let it thunder to the tune of Green Sleeves, hail kissing-comfits and snow eringoes; let there come a tempest of provocation." If we could all go back to that, back to the potato—or indeed to Viagra, which has been a great boon to endangered species—how much would be saved.

The majority of hermit crabs are asymmetrical; they have

ten legs, but the front left claw is enlarged for defense, and the front right is smaller, for scooping food, about which they aren't fussy: algae, plant life, other dead crabs. They have, too, under their shells, rear ends that twist in on themselves—helter-skelters. And they're off-kilter beautiful: the jeweled anemone crab has shocking emerald eyes, on stalks that are striped like a barber's pole in red and white. They can be sea-gray or royal purple; the giant spotted hermit crab is orange with white dots edged in black; the hairy yellow crab is striped yellow and cream, with opulent hairs on its legs and eyes on blue stalks. Up close, even the coconut crabs are beautiful: some are aquamarine at the hinges, some rich brown with a burnt-orange back. A teeming horrorscape, but make it fashion.

Hermit crabs can, if they must, make their home almost anywhere. They have been found in tin cans, in coconut halves. The Pylochelidae family evolved to make their homes not in shells but in sea sponges, stones, driftwood, pieces of bamboo. More and more, in these darker days, I admire resourcefulness. I love their tenacity: forging lives from the shells of the dead, making homes from the debris that the world, in its chaos, has left out for them.

THE
Seal

There was a time when a person walking past a specific pond in Maine might be catcalled by a seal: "Come over here!" in a thick Maine accent. His name was Hoover, and he had been adopted as an orphaned pup by a lobster fisherman named George Swallow in the 1970s, and he spoke English. "What are ya doin'?" and "Hello, hello there!" George spoke to the seal constantly, calling him, cherishing him: Hoover, so christened because he hoovered up fish, would barrel into him in the morning, buffeting his face with "kisses." Eventually, when his fish needs became too expensive for the Swallows, he was given to an aquarium. Swallow told the attendants that Hoover could speak, but, on being met with raised eyebrows and skeptical faces, let it drop. Hoover, stunned by the new environment, remained silent for several years, but when he finally spoke again, it was the start of a vocal streak that lasted the rest of his life. He could never be persuaded to talk on command, but there is a recording of him, calling in a guttural voice: *hello, hello, get over here*. Seals, it transpired, have surprising language-learning capacity—scientists at St. Andrews University in Scotland have taught their charges to sing "Twinkle, Twinkle, Little Star."

Even those who do not speak can sound human when they cry. In *Moby-Dick*, published in 1851, the ship sails over water from which rises "a cry so plaintively wild and unearthly" that the crew freezes, transfixed. "The Christian or civilized part of the crew said it was mermaids, and shuddered . . . yet the grey Manxman—the oldest mariner of all—declared that the wild thrilling sounds that were heard, were the voices of newly drowned men in the sea." Only Ahab gives a hollow laugh, "and thus explained the wonder":

> some young seals that had lost their dams, or some dams that had lost their cubs, must have risen nigh the ship and kept company with her, crying and sobbing with their human sort of wail. But this only the more affected some of [the men], because most mariners cherish a very superstitious feeling about seals, arising not only from their peculiar tones when in distress, but also from the human look of their round heads and semi-intelligent faces, seen peeringly uprising from the water alongside.

Ahab tells: "In the sea, under certain circumstances, seals have more than once been mistaken for men."

Hoover was a harbor seal, one of the most common of the world's thirty-three species of seal—formally, we should call them pinnipeds, from the Latin *pinna*, fin, and *pes*, foot. The majority of these species live in Arctic and Antarctic waters; among them the harp seal, which in adulthood can be gray, or speckled, or silver: not the polite "silver" which is actually the color of dishwater, but true burnished silver. The pups are born on ice, stained lurid yellow with amniotic fluid, but once cleaned they are snow-white until the first molt. They have round black eyes which, were they human, would slay the whole of Hollywood. Seals are generally impressive mothers, careful and valiant and liable to bite if you get in the way;

but the mothering of the harp seals takes place, increasingly, in a race against time. As soon as the pup is born, a countdown begins to get it weaned and ready to swim before the ice melts. To that end, the mother's milk is fifty percent fat, the consistency of mayonnaise (the richest ice cream you're liable to encounter is fifteen percent; human milk is four percent), and pups can double their body weight within days. The mother then takes the pup into the water, and uses her belly as a float to help her baby rest. It takes only minutes to learn: to go from flailing and thin panicked piping to swimming. But climate change, disrupting the breath and flow of the ice, has made the pups' survival harder. For the last thirty years, we have lost more than thirteen percent of the Arctic's sea ice each decade. Ice reflects the sun's heat back upon itself, stabilizing the climate, where open water absorbs it. In 2017 on the Gulf of St. Lawrence, the ice broke so early that a harp colony's entire pup population drowned overnight: so much work to produce so much life, dead. As the ice melts, there will be fewer and fewer places for colonies to go: a whittling away of refuge.

The seal family's strangeness is as monumental as their beauty: male hooded seals, for instance, have a nasal cavity which allows them to blow up what looks like a red fleshy balloon the size of a football from their noses, in a bid to scare off rivals looking to mate with their partners. The world's heaviest carnivoran, the elephant seal, weighs up to 8,800 pounds, about the same as a flatbed truck, and though on beaches and bays they have all the elegance of a landslide, underwater they move with the confidence and clarity of athletes. Though no seal can breathe underwater, elephant seals can dive more than a mile deep, and high levels of myoglobin, aiding the storage of oxygen in the muscles, allow them to hold their breath for up to two hours. Some have a flair which is less Hollywood, more character actor; bearded seals have a copious crop of white whiskers against gray, making both male and female seals look like elder statesmen when wet, or, when the whiskers are dry and curl upward, like a clan of rakish musketeers. Others would be the envy of the abstractionists; ribbon seals are ringed with thick geometric white circles and stripes, ambulant Malevich paintings; Mediterranean monk seals birth pups that are coal black, with painterly sweeps of white across the belly—though for how much longer is in the balance. The global population currently stands at a few hundred individuals.

With their faces so watchful, so capable of pathos and mischief, it makes sense that Early Nordic sagas showed an ambivalence about what they did and didn't know, could and couldn't do. In the thirteenth-century *Laxdæla Saga*, a warrior, Þóroddr, is sailing with his crew to claim new land with an overburdened boat, when "they saw a seal, much larger than most, swimming in the water nearby. It swam round and round the ship, its flippers unusually long, and everyone aboard was struck by its eyes, which were like those of a human." The warriors attempt to kill it, but it evades them. It watches as, moments later, "a great storm struck which capsized the ship. Everyone aboard was drowned except one man." The seal had either foretold, or caused, their death.

Most elusive and eldritch of all is the selkie. A shape-shifter between human and seal, it haunts sea lore. In the most common stories, found from Orkney and Shetland to Iceland and the Faroe Islands, the selkie sheds her seal skin on the shore to walk naked on the land. A man, enamored of her sleek loveliness, steals the skin and so forces her to become his wife. They have children, but she weeps and longs always for the ocean, and as soon as she can find the skin, she puts it on and escapes to the water, joyfully abandoning her children in favor of the sea. Other selkie stories are the reverse: of Shetland selkie women luring landsmen into the waves, who never again return to solid ground. Male selkies are exquisite in their human form, female selkies staggering in their beauty: they cannot be resisted, for they have in them the power of the sea. The ballad "The Grey Selkie of Sule Skerry" tells the story of a woman who, lamenting she does not know where her baby's father is, finds him suddenly rising from the sea foam to claim it:

I am a man upo' da land;
I am a selkie i' da sea.
An' when I'm far fa every strand,
My dwelling is in Shöol Skerry.

Once, while I was swimming in the North Sea off Stiffkey in Norfolk, a herd of harbor seals rose in the water. They did not retreat—a few advanced. To be in ice-cold water, under steady gray skies with their gray steady beauty: it felt like being churched. It is very easy in their presence to understand why we have cast them as singing, knowing, watchful things.

THE
Bear

The Renaissance poet John Donne had a theory about bears: a bear cub, he believed, is born a solid lump of flesh, until its mother bites and licks it into shape. The idea can be traced to Pliny's *Historia Naturalis*:

> . . . *at first, they seem to be a lump of white flesh without all form, little bigger than rattons, without eyes, and wanting hair: only there is some show and appearance of claws that put forth. This rude lump, with licking they fashion by little and little into some shape: and nothing is more rare to be seen in the world, than a she bear bringing forth her young.*

The image appears several times in Donne's work: most memorably, as a warning—that we must not, in the fever of our love, devour chunks of each other:

> *Love's a bear-whelp born: if we o'er-lick*
> *Our love, and force it new strange shapes to take,*
> *We err, and of a lump a monster make.*

He wasn't alone in loving the image: the playwright George Chapman used it, too, to describe a half-formed bad idea: "Nay, I think he has not licked his whelp into full shape yet"—and Shakespeare used the same image of a bear-whelp in *Henry VI, Part 3*—Gloucester is "like to a chaos, or an unlick'd bear-whelp." It was the seventeenth-century polymath killjoy Thomas Browne who debunked the theory, in his *Pseudodoxia*, "that a Bear brings forth her young informous and unshapen, which she fashioneth after by licking them over, is an opinion not only vulgar, and common with us at present: but hath been of old delivered by ancient Writers." Browne offered a rational explanation for the myth:

> Now as the opinion is repugnant both unto sense and Reason, so hath it probably been occasioned from some slight ground in either. Thus in regard the Cub comes forth involved in the Chorion, a thick and tough Membrane obscuring the formation, and which the Dam doth after bite and tear asunder; the beholder at first sight conceives it a rude and informous lump of flesh, and imputes the ensuing shape unto the Mouthing of the Dam.

Since long before the first teddy bear was created in 1902, we have wanted bears around us: in their vast bulk, their beauty, and their teeth. In 1251, Henry III was given a "white bear" by the king of Norway—a polar bear, which was kept in the Tower and allowed to hunt for fish in the Thames. In 1609, an English expedition to the Arctic came across "a she-bear and two young ones: Master Thomas Welden shot and killed her. After she was slain, we got the young ones, and brought them home, into England, where they are alive in Paris Garden." They were put in the charge of Edward Alleyn, one of the great actors of the age, who doubled as the Master of the Royal Bears. On New Year's Day 1611, Ben Jonson's *Oberon, the Faery Prince* was performed before the king: a sixteen-year-old Prince

Henry, pale-faced and exquisitely dressed, rode a chariot onto the stage at Whitehall, driving before him the two polar bear cubs. But none lived as closely with their bears as Lord Byron, who, furious that the statutes at Trinity College Cambridge forbade him to have a dog in his rooms, bought a bear. He wrote to a friend in 1807: "I have got a new friend, the finest in the world, a tame bear. When I brought him here, they asked me what to do with him, and my reply was, 'he should sit for a fellowship.'"

They are a species of large and capacious astonishments. The Kodiak brown bear is a prodigy of growth: at birth, a cub weighs less than a pound—a small loaf of bread—but can end by weighing more than 1,500 pounds; if we grew at the same rate, an adult man would weigh the same as a rhinoceros. Their sense of smell is a hundred times better than ours: a polar bear can smell you more than eighteen miles off, and swim one hundred miles without stopping to rest: England to France five times over, without pause. And for those species which hibernate, they can unfurl a powerful slowness within themselves and pass more than a hundred days without eating, drinking, or urinating, their heart rates dropping from forty beats per minute to just eight slow beats, drawing one breath every forty-five seconds. More astonishing: their bodies become recycling plants, converting urea into protein, and a neat plug forms in the anus, probably to stop them from defecating in the den.

The bear with the finest origin myth is, perhaps, the panda.

The story is told to children in both China and Tibet: that long ago a shepherdess, guarding her sheep, was joined every day by a panda cub. Back in those early days, all pandas were snow-white, and possibly the cub thought the sheep were co-pandas. One day, as the panda cub was gamboling clumsily with the lambs, a leopard attacked it. The shepherdess threw herself in front of the panda, and was killed. The panda cub and his family came in somber gratitude to the shepherdess's funeral, and out of respect they covered their arms with black ashes, as was the custom. As the funeral went on, they wept, wiping their eyes with their paws and staining them black. As their weeping grew louder, they covered their ears, so that they would not have to hear their own sobs. The ash never washed off, and therefore they are forever marked with signs of their love and grief, their ever-ongoing fealty to bravery.

Bears are so vast that we've used them to make ourselves feel pleasantly conquerous. Down by the river in Elizabethan England, alongside the brothels—some very grand with moats and flags, some much less so—and theaters, peanut sellers and pubs, there were the bear pits. A bear would be chained by its leg or neck to a stake in the middle of the ring, and a pack of dogs set on it. One of Elizabeth I's courtiers was an eager attendee: "it was a very pleasant sport to see . . . the bear, with his pink eyes, tearing after his enemies' approach . . . with biting, with clawing, with roaring, with tossing and tumbling, he would work and wind himself from them. And when he was loose . . . shake his ears twice or thrice with the blood and the slather hanging about his physiognomy." The bears were rarely baited to the death—they were too valuable to lose—and instead they fought over and over, becoming known across London: "Harry of Tame," "Sampson," and a she-bear called "Boss." But there were exceptions to the death rule: the year before his polar-bear chariot, Prince Henry went to the Tower with his parents to see "a great fierce bear, which had killed a child that was negligently left in the bear-house." The king wanted to see it pitted against his collection of lions: but the lions, terrified, refused to fight. The bear was instead handed over to mastiffs and torn to pieces, and the mother of the dead child was given £20 out of the spectators' entrance fees.

In the eighteenth century, bears met another kind of human hunger: our burning-hot desire for beauty. Bear grease was the luxury cosmetic of the day: the Chanel No. 5, the Vidal Sassoon. It was said to be a cure for baldness, and it added a glistening sheen to the wigs of the upper classes. Most bear fat was really just pig fat, dyed green—but some traders had bears in cages outside their shops to prove their bona fides. The richest customers would gather to watch the fat taken straight from the carcass, as a guarantee they were getting the authentic product. One enterprising barber claimed to house forty live bears in his cellar for the purpose. The noise, and smell, must have made visits to his premises a not unadulterated pleasure.

A question children like to ask: Who would win in a fight, man or bear? It depends on the day. There are about forty attacks by bears reported globally per year, but fewer than twenty percent of them are fatal. (Far, far fewer than the number of people killed yearly by falling televisions, faulty lawn mowers, or toppling vending machines.) Sometimes neither party wins. In 1883, a Kansas newspaper reported:

> *The body of Frank Devereaux was recently found in the woods eight miles from Cheboygan, Michigan. The surroundings show that he was killed in a bear fight, which resulted fatally for both, as the animal's body was found near that of the dead man. The body was terribly cut up in the contest, and the ground torn for a space of twenty feet, showing that the struggle had been a fearful one.*

The traditional advice runs: if a brown bear charges you, play dead; if a black bear, look as large as you can and roar; in both cases, never run away. There's a rhyme: bear brown, lie down / bear black, fight back. It is advice that requires exceptional presence of mind in the presence of teeth.

In the wider fight between bears and humans, though— the one that absolutely none of us meant to enter into, but in which we are all engaged—we are winning. Of the eight species of bear, six are at risk or endangered: the American black bear and the brown bear are coping well enough, but not the Asian black bear, the sloth bear, the spectacled Andean bear, the sun bear, or the polar bear, nor the giant panda, with its limbs like slow and ungainly poetry in motion. Sloth bears are tortured into dancing, trained up on heated slabs of metal as cubs, their feet smeared with Vaseline; polar bears stand atop melting ice. Thousands of Asiatic black bears and sun bears are currently in cages, milked for their bile in both legal and illegal farms. Bear bile, which is used to dissolve gallstones (at which, unlike many traditional remedies, it is effective) and to help with fever and cleanse the liver, is a trade worth $2 billion. It is hard to remember how urgently something needs protecting when it could also kill you.

The best stage direction ever written is Shakespeare's, in *The Winter's Tale:* "Exit, pursued by a bear." Was it a real bear on the stage? Probably not: just capacious roaring offstage, or actors clad in bearskins. But the bears were nearby, listening, waiting for the dogs that we would set on them in our desire to see something so large, so fierce, and so beautiful stand and roar.

THE

Narwhal

In 1584, as Ivan the Terrible lay dying, he called from his bed for his unicorn horn, a royal staff "garnished with verie fare diamondes, rubies, saphiers, emeralls." Unicorn horns were believed throughout Europe to have magical curative properties; as late as 1789, a unicorn drinking horn was used to protect the French court, where it was said to sweat and change color in the presence of poison. To prove the horn's efficacy, Ivan ordered his physician to scratch a circle on the table with the tip of the horn, and to "seeke owt for som spiders." The spiders placed within the circle curled up and died; spiders placed outside it ran away and survived. The dead spiders, though, could not console Ivan. "It is too laite," he said, "it will not preserve me." Whereupon he died.

The unicorn horn was, of course, a narwhal tusk: the tooth of a small Arctic whale, which grows out through the upper lip, twisting counterclockwise for up to eight feet. Named rather ungallantly for the Old Norse word *nar*, meaning "corpse," and *hvalr*, meaning "whale," after their mottled gray markings, narwhals are unicorn-like not just in their appendages, but in their elusiveness; they are one of the mammals about which we know the least. They spend the winter months dodging dense pack ice, where humans cannot follow, and can swim a mile deep, twisting upside down as they descend into pitch-black water.

The great mystery of the narwhal is the purpose of its tusk. Appearing when the calf is about a year old, as short and thin as a little finger, it grows for nearly ten years, until it's as wide as ten inches at the base. Herman Melville writes of the "nostril whale" in *Moby-Dick:* "Some sailors tell me that the Narwhale employs it for a rake in turning over the bottom of the sea for food. Charley Coffin said it was used for an ice-piercer . . . But you cannot prove either of these surmises to be correct." He ends by suggesting it would make an excellent letter opener. Because fewer than fifteen percent of female narwhals have the tusk, it can't be necessary for survival, and so, when male narwhals were observed clashing tusks, it was often interpreted as rivalrous jousting. Recently, though, scientists have found that the tusk is shot through with around ten million nerve endings, and by rubbing tusks on meeting, the narwhals may be passing on information about the salinity (and therefore propensity to freeze) of the water through which they have just passed: not aggressors, then, but cartographers. Others have been spotted using the horn as a kind of piscine weapon, to stun fish in the water before eating them. The horns may also be an aid to courtship; a positive correlation has been discovered between testicle size and horn length, and the tusk may be a way for the most fertile male narwhals to advertise themselves.

The narwhal is exquisitely formed. To conserve heat, the surface area of its skin is as streamlined as possible: no ears, no lips, no eyelashes, no inconveniently extruding sexual organs; nothing to hold back the swift passage through water. As much as forty percent of the narwhal's body mass can be made up of blubber, allowing it to keep a warm mammalian body temperature amid the dark floating ice. Narwhals mate in a kind of ballet; a pair will swim alongside each other for hours, skins touching, until the female twists belly up to press her body against the male. Later, when the female narwhal gives birth, an adolescent female from the pod will frequently act as nursemaid, swimming beside the mother with the calf between them to create a current that sweeps the baby along, the whale equivalent of a papoose.

The legend of the narwhal is not a gentle one. The Danish-Inuit ethnologist Knud Rasmussen recorded the myths of the Inuit of Greenland's northwestern coast in the early twentieth century. In the narwhal origin myth, the cruel mother of a blind son tricks him out of his fair share of bear meat. The mother plaits and twists her hair into a long braid and the two go out to harvest passing white whales; the son, in revenge for her cruelty, binds her with ropes to one of the whales, and it drags her into the sea. According to Rasmussen, "she did not come back, and was changed into a narwhal . . . and from her the narwhals are descended."

One of the earliest written accounts of the narwhal dates from 1577. Martin Frobisher, seaman and privateer, led an expedition to Baffin Island, where his men discovered a dead narwhal on the beach. They tested it for magic, using the same method as Ivan the Terrible:

On this West shoare we found a dead fishe floating, whiche had in his nose a horne streight and torquet, of lengthe two yardes lacking two ynches, being broken in the top, where we might perceiue it hollowe, into which some of our Saylers putting Spiders, they presently dyed. I sawe not the tryall hereof, but it was reported vnto me of a trueth: by the vertue whereof, we supposed it to be a sea Unicorne.

Triumphantly, they took the horn. They also forcibly took three Inuit from their homes: a man, Calichough; a woman, Egnock; and her child, Nutioc. All of them died soon after arriving in England.

"When Sir Martin returned from that voyage," we're told, "he presented to her highness a prodigious long horn of the Narwhale, which for a long period after hung in the castle at Windsor." This was not Elizabeth I's only narwhal tusk. Sir Humphrey Gilbert, Walter Raleigh's half brother, presented her with a gem-encrusted narwhal tusk worth £10,000 (enough, at the time, to buy and staff a small castle). It was, he told her, a "sea-unicorn." Gilbert's Latin motto was *Quid non* ("Why not?"), but in this instance he probably believed himself to be truthful; unicorns, after all, appear nine times in the Bible. It wasn't uncommon for the richest and finest churches to have a unicorn horn, from which they would slice pieces and mix them in holy water to help their ailing parishioners—Chester Cathedral in England has a seventeenth-century narwhal tusk, offering its silent magic up to worshippers.

Narwhals are designated "near threatened." The greatest threat to their survival is climate change, which is shrinking the ice cover too quickly for them to adapt; without ice cover they will have nowhere to hide from killer whales, nowhere to feed. Narwhals communicate via a series of clicks and buzzes (higher in pitch than a humpback whale's, less shrill than a dolphin's); with increased shipping and industrial extraction in the Arctic, noise pollution risks rendering them inaudible and effectively mute, and thereby unable to protect and teach their young—we have taken their silence and replaced it with a nightclub roar. For now, though, there are perhaps eighty thousand in existence. In some corner of the sunless sea, passing through waters cold and dark enough to keep us at bay, there moves a beauty and a strangeness that rivals the unicorn.

THE
Crow

I f you were to choose an animal to double-cross, it would be wisest not to choose a crow. They're formidable in their intelligence: they are clever and wise enough to hold grudges against us. For five years, students at the University of Washington wearing caveman masks hunted down and captured crows that lived in the trees on campus, held them briefly in captivity, and then released them. The crows, like very small Old Testament gods, did not forget, nor were they indiscriminate in their wrath. When the students walked past below without the masks, the crows ignored them, but when they wore caveman faces, the crows mobbed them, scolding and screaming at them. The fury and fear were passed, from one crow to another, through the group: in another experiment, mask-wearing students were attacked and reprimanded after the original crows were dead.

But if they make fierce enemies, they make even finer allies. A girl in Seattle called Gabi Mann made worldwide news when the crows she had fed every day since she was four years old began to bring her gifts in return: a paper clip, a blue bead, a piece of Lego, a tiny silver heart from a pendant. But, even better: Her mother, Lisa, dropped a camera lens cap while out taking photographs in a field. The crows watched nearby. She was almost home before she realized it was lost, but as she came down her garden path she saw it had been returned to her, balanced precisely on the rim of the birdbath. Camera footage showed a crow arriving with it, walking it to the birdbath, washing it several times over, and laying it out to wait for her return. They notice us; they punish and reward.

Almost all birds are builders, of course, but few are such fine artisans. The crow is an Einstein among birds, their brain to body mass ratio only a little lower than our own. Many can fashion tools, snapping twigs from trees, stripping them, bending them into hooks and then digging into small spaces for food. If the tool is a good one, they stash it away for use later; some have been seen to pilfer the most prized utensils from each other. Wild New Caledonian crows native to the Pacific archipelago have been taught, with ease, to use vending machines. At the Puy du Fou theme park in France, six particularly intelligent midnight-blue rooks—members of the crow family—have been trained to pick up litter. They drop cigarette butts in a box, which releases a small piece of food, while the people who dropped the butt look on. Imagine it: being publicly shamed by a rook.

The crow family is a large one, clad in many dozen shades of finest black. The genus *Corvus* includes crows, ravens (which are distinguished from crows only by their size), and rooks, which can be told apart by their silver-gray beaks—while the wider Corvidae family, of 133 species, includes magpies, jays, and jackdaws. They're not the sweetest family: they have a blood-thirsty edge to them. Crows have been known to peck the eyes from weak newborn lambs. Magpies eat the eggs and young of sparrows and starlings and are consequently rampantly unpopular in parts of the countryside, such that the magazine

Sporting Shooter once offered a £500 prize to whoever could produce the most magpie corpses. (Those who blame magpies for the catastrophic fall in numbers of songbirds across Europe, though, are laying the burden in the wrong place; most of it is due to pesticides, and to our new practice of farming winter-sown cereals, which leave the crops too high and dense for the birds to nest in.) The crow family has gone largely unloved through history, in part because of the grisly lamb-time feasting, in part because their eyes are so knowing and skeptical, and in part because their voices are, generally, guttural and gravelly things—witness the raven who drove Edgar Allan Poe to demand "What this grim, ungainly, ghastly, gaunt, and ominous bird of yore / Meant in croaking 'Nevermore.' "

There is, though, a crow with a voice to provoke joy. Its versatility is astonishing; some of its calls are musical, some nearly human: a one-bird cornucopia, the Hawaiian crow, also called the 'alalā. It's a large bird, exquisitely petrol-blue-black and exuberantly declarative. One call sounds like a whistling kettle. One sounds exactly like Elvis's yelp, and is listed on the U.S. Fish and Wildlife Service's delineation as the "yeeow." They were once one of the most numerous birds in Hawaii, calling through the forests in rattles, growls, and fantastical wails.

In Hawaiian folklore, the 'alalā is one of the guardians of the soul. The soul of the dead travels to a Leaping Place—usually a promontory high above the ocean—to await its guide to its final rest. For those who die on Ka'ū on Hawaii island, one of the Leaping Places is the cliff at the southern tip of the island,

Ka Lae, and the guide is the crow; there the soul and bird meet, and together they leap into the afterlife. Without the bird, the soul risks becoming lost, wandering forever among ghosts and night moths.

The 'alalā was declared extinct in the wild in 2002. In 2016, thirty were reintroduced to the forests. Of those thirty, five survived, and were swiftly taken back into captivity. Some died because they had no anti-predator response—their prime threat from the air is the Hawaiian hawk, the 'io, which is itself endangered. The forests in which they used to breed, shelter, and yelp have been denuded by cattle and human expansion; and, once the birds were gone, dozens of native Hawaiian plants which relied on them for seed dispersal also began to vanish. The attempt to reintroduce them into the wild is well funded and ongoing, but the odds are hard. Male 'alalā, reared in captivity, are more aggressive, prone to attack, and less likely to understand when their mates are making romantically encouraging gestures. Inbreeding has made their egg shells thinner and their clutches smaller. Though they may enter the wild again, there will be no gossiping, shrieking birds to teach them how to be true wild crows—they will be different. It is only one species, of only one bird. But they could use tools, and they could call to each other in ways more sophisticated than we could yet decode; it is so much intelligence wiped out. And if the 'alalā are not saved, one of the ways in which humans have painstakingly and generously explained death to one another will be dead, and there will be no guides awaiting the souls at Ka Lae.

THE
Hare

Hares have always been thought magic. In their long-limbed, quivering beauty, they were believed to be walking, breathing love potions: charms born ready-made. The Greek sophist Philostratus warned his third-century readers that there were unscrupulous men out there who had found in the hare "a certain power to produce love, and try to secure the objects of their affection by the compulsion of magic art." Pliny suggested eating hares would increase your glamour: "the people think that if you eat a hare your body acquires sexual attractiveness for nine days, a vulgar superstition, which however, must have some truth in it since the belief in it is so widespread." Martial, father of the epigram, writes to a noblewoman named Gellia, rather acidly: "Whenever you send me a hare, Gellia, you say to me, 'You will be beautiful for seven days.' . . . my love, if you mean what you say, *you*, Gellia, have never had a hare to eat." The hare was sacred to Aphrodite, goddess of love, and Eros can be found on droves of Greek vases, pursuing the hare across the pottery or cradling one in his arms.

This sense of the hare as belonging to the arena of sex and desire stems in part from a belief in its astonishing fertility. Aristotle suggested in his *Historia Animalium* that the hare could get pregnant twice: they "breed and bear at all seasons, *superfoetate* [conceive again] during pregnancy." Aristotle also suggested, of course, that eels spontaneously generate from the mud: but in fact he was right—the hare can get pregnant while already pregnant. A male hare can fertilize a female during the latter part of her pregnancy: the embryos will begin to develop while they wait in the oviduct until the delivery of the first pregnancy, and then, as soon as the uterus is free, move in. The time saved means they're able to deliver around thirty percent more offspring during a breeding season; and perhaps because of the sense that they could control their conceiving with such finesse, it was long thought that hare parts could act as contraception. Aetius, offering medical advice in the imperial court in Constantinople in the sixth century, suggested that women "take seeds of henbane [a plant known more commonly as stinking nightshade], mix the milk of a she ass, a little myrtle and a berry of black ivy . . . to be worn after having been wrapped in the skin of a hare." (This was the same man who advised inserting children's milk teeth into a woman's anus as an alternative method of contraception, advice one would so profoundly hope nobody took.)

Part of its magic was in the belief that it was a switcher-at-will between two lives, hermaphrodite. It was, the fourth-century rhetorician Donatus writes: *"modo mas, modo femina"*: sometimes male, sometimes female. Because of this, it became in antiquity a way to express homosexual love: in a comedy by the Roman African playwright Terence, one bold young man is told by another: "What are you saying, impudent creature? You're surely a hare and you seek flesh." On a clay pot from the year 500 BCE, a man stands in love with a youth; he offers the younger man a live hare, and the younger man gazes at it. It was a thing to conjure love, both hetero- and homosexual: a thing, which, laid at your feet, alive, in its fine-featured loveliness, would sprint away your doubts.

Their name varies, place to place: in America, they are jackrabbits. Mark Twain made the name famous in *Roughing It*, his "record of several years of variegated vagabondizing." The "jackass rabbit," he writes, "has the most preposterous ears that ever were mounted on any creature *but* a jackass. When he is sitting quiet, thinking about his sins . . . his majestic ears project above him conspicuously." With time and familiarity, they became jackrabbits. The species' Latin name, *lepus*, came, the ancient Roman scholars said, from the Latin *lavipes*, light foot: and my God they deserve it. Brown hares can run at almost fifty miles an hour and jump almost ten feet: five times their own length in a single bound. Twain, an admirer of speed, wrote that the animal "scatters miles behind him with an easy indifference that is enchanting." To see a hare outrun a fox, zigzagging to disrupt its predator's momentum, is to know you are in the presence of the marvelous. Leverets—young hares—are born with their eyes open and fur on, ready to sprint: they live largely alone and do not make burrows, but rest in shallow indentations in the ground, never halting, always moving. "To kiss the hare's foot" is to be too late for dinner, according to *Brewer's Dictionary:* "the hare has run away, and you are only in time to 'kiss' the print of his foot."

But they are not fast enough, of course, to outrun us. In Mexico, the Tehuantepec jackrabbit, which wears white trousers and a black stripe from ear to nape, is critically endangered, and still depleting. The number of hares in Britain has declined by eighty percent in the last century, in part because the relentless destruction of hedgerows killed their cover: between 1985 and 1997, more than a quarter of the almost 400,000 miles of hedgerows that ran through England and Wales were torn out and burned. (Those hedgerows that still exist are largely at the mercy of British whims and fashions and carefulness: and it takes only a very little reading of history to lose confidence in that particular trinity. There is a stretch of hedgerow in Cambridgeshire—Judith's Hedge—that is older than Durham Cathedral, older than St. James's or Buckingham Palace: it has spent nine hundred green and thorny years hosting life within it. Anyone seeking to tear it down would have to apply to the local council—but many younger hedgerows have no such protection. A proposition: those who have profited from the destruction of such places should be made to live in a motorway service station their whole lives.) In America, the laws on hunting vary from state to state—in Missouri jackrabbits are protected, but in Utah you can shoot to your heart's content. In England, hares are the only game you can legally kill all year round. The English hunt them even in their mating season, when the females are boxing at would-be suitors in the fields, and shoot them in hordes: as many as three hundred thousand a year. We might remind each other: we are perhaps slaughtering the Easter Bunny. He was a hare long before he was a rabbit; the hare was, country gossip goes (probably a fact more hopeful than technically historical), sacred to Eostre, the Saxon goddess of spring—no rabidly cute bundle of fluff, then, but Eostre Hare.

Some hares were also witches, fairy folk. In Elizabeth Goudge's astonishing children's book of 1946, *The Little White Horse*—a book which, written just as the war ended, revels in sugar biscuits and exquisite beauties—the hare is the finest beauty of all. They are not, a boy explains, like rabbits: "a hare, now, is a different thing altogether. A hare is not a pet but a person. Hares are clever and brave and loving, and they have fairy blood in them." One Celtic legend has the warrior-poet Oisín hunting a hare, injuring it in the leg: he follows it into thick brambles, and, finding a door down into the ground, comes into a large hall, where sits a lovely woman, bleeding from a leg wound. The story sounds closely akin to a real-life accusation: a trial for witchcraft in 1663 of an old woman called Julian Cox. A witness at the trial stated:

> *A huntsman swore that he went out with a pack of hounds to hunt a hare, and not far from Julian Cox's house he at last started a hare: the dogs hunted her very close . . . till at last the huntsman perceiving the hare almost spent and making towards a great bush, he ran on the other side of the bush to take her up and preserve her from the dogs; but as soon as he laid hands on her it proved to be Julian Cox . . . He knowing her, was so affrighted that his hair on his head stood on end; and yet he spake to her and ask'd her what brought her there; but she was so far out of*

breath that she could not make him any answer . . . the hunts-
man and his dogs went home presently sadly affrighted.

So they were dangerous, as well as beautiful. A book on folklore from 1875 told that the hare moved in close association with calamity—it was recommended that, passing a hare, you should recite: "Hare before, Trouble behind: Change ye, Cross, and free me." A Middle English poem, "The Name of the Hare," offers the creature's seventy-seven noms de plume, which you should say if you pass one, to ward off ill luck: few are complimentary. Seamus Heaney's translation has, among others:

The creep-along, the sitter-still,
the pintail, the ring-the-hill,
the sudden start,
the shake-the-heart,
the belly-white,
the lambs-in-flight.

The gobshite, the gum-sucker,
the scare-the-man, the faith-breaker,
the snuff-the-ground, the baldy skull,
(his chief name is scoundrel.) ["His hei nome is srewart."]

The gobshite and faith-breaker, certainly, for farmers: but they have been thought holy, too. The "three hare" motif appears across holy spaces in the Far East and in churches throughout Europe: three hares running in a circle, their ears shared and entwined. Hares have long been paired with the Virgin in her portraits. (We have, perversely, thought them the epitome both of sex and of virginity: it was for a while thought that their fertility was so profound they could reproduce without a mate.) Their running bodies, set up upon the roof bosses of churches, call up the Holy Trinity: One in Three, Three in One, prime mover in constant motion.

If beauty is enough to merit love—and we have historically reckoned that it is—then we should love the hare more than almost any other creature. The closer you get, the more beautiful; of the thirty-two species, some, like the Burmese hare, are reddish-gray shading to silver, others, like the Tibetan woolly, the color of just-cut straw; the Indian hare wears a patch of black at the back of its neck like a hair ribbon. The mountain hares in the north of Britain turn stark white in winter: real white, because the white only lasts a season, not the grubby white of most habitually white creatures. Their legs have the grandeur of Olympians, and their ears, black-tipped, lined in pink velvet and thin enough to be semi-translucent in the light, dwarf those of a rabbit. Seen in flight, the ears are banners across a battleground, never furled. "The cat of the wood," Heaney calls them: "the stag of the cabbages." If there is magic in this world, some part of it lies with them. So if you are reading this, my love, I don't need flowers, or jewels. Please, bring me a hare.

THE

Wolf

Each week throughout the seventeenth century, a spreadsheet was drawn up to record the causes of death in London, called the Bill of Mortality. The causes listed are vivid, and raise many questions: "Affrighted," "Blasted," "Teeth," "Dead in Street," "Eaten of Lice," "King's Evil." In one 1650 account, eight cases read: "Wolf." It's so tempting to imagine a fanged shadow prowling past Buckingham Palace, but in fact "wolf" was the name given to a far deadlier killer. In 1615, a clergyman wrote of "disease in the breast, call'd the Cancer, vulgarly the wolf." In 1710, a translation of the writings of the French surgeon Pierre Dionis read " 'Tis a Disease which attacks not only the Breast, but several other Parts, on which it is not less outrageous. It sometimes assumes different names; when it comes on the Legs, 'tis called the Wolf, because if left to itself, 'twill not quit them 'till it has devoured them."

The "wolf" moniker developed a bite all its own: the link between wolves and cancer became so entrenched in the popular imagination that in 1714, the physician Daniel Turner wrote of "a famous Cancer Doctor" who claimed to have cured a woman's cancerous ulcer: "Such an [tall tale] I was not long since inform'd of, by a Woman who vowed that . . . when they held a Piece of raw Flesh at a Distance from the Sore, the Wolf peeps out, discovering his Head, and gaping to receive it." The image—of an actual wolf peeking out from a woman's flesh like a fairground whack-a-mole—shows, for all its lunacy, the potency of our metaphors: we start to believe them. In 1599,

The Boock of Physicke suggested that a cure for cancer was eating dried and powdered "wolves-tunge"; our figurative language has fairy-tale power over us, possessing us as we conjure with it.

We decided very early on that wolves are deceptive, ravenous, and morally backward: duplicitous as well as hungry. It was biblical: the first use in English of the idea of the wolf as rapacious comes in 950, in a translation of the Lindisfarne Gospels: *"Heonu ic sendo iuih suæ scip in middum vel inmong uulfa"* ("Behold, I am sending you out as sheep in the midst of wolves"). When William Caxton, the man who introduced the printing press to England, published Aesop's fables in 1483, he chose three wolf stories. And it made sense: wolves were a scourge: particularly if you were, or owned, a sheep. The Norman kings, ruling in the eleventh and twelfth centuries, had servants whose specific role was to hunt down wolves; felons could be spared death if they agreed to become wolf-stalkers, and the animals wreaked such havoc with livestock that Edward I went so far as to order the extermination of all English wolves. He was largely successful: the last reference to wolves in England seems to be of one marauding through a deer park in 1290. By 1300, an English doctor was hauled up in front of the medieval equivalent of customs officers when he attempted to import the bodies of "four putrid wolves" into the country for medical research: they could not, by then, be found at home.

Wolves were once found across the entirety of North America, from the Florida black wolf stalking through what would now be Disney World, to the snow-white Alaskan tundra wolf. Over several hundred years, settlers hunted them almost to extinction, and by the 1950s, America was almost emptied of wolves—the Florida black, for instance, is gone forever, and the red wolf remains critically endangered. But in 1973 the U.S. Endangered Species Act halted the decline, and their numbers began to rise. Now their fate yo-yos, resting entirely in human hands: in 2019, the Trump administration lifted federal protections for gray wolves, leading to an immediate surge in hunting. In Wisconsin's first wolf hunt, 218 wolves were shot in three days, twenty percent of the state's entire wolf population—but the wolf's protections have, since 2022, been restored. How long they will last remains uncertain: if wolves are reputed to be insatiable and capricious, so are we.

In fact, they don't deserve that reputation. They certainly have, across the world, eaten our livestock, and, back when shepherds were often children and wolves were emboldened by rabies, they have occasionally eaten our offspring, but they are no more wily and evil than lions, tigers, bears. They are simply medium-large predators: but we have refused to see them with steady eyes. We needed a symbol into which to pour our fear and mistrust of the world, and we have chosen the wolf, and chosen it with passion and commitment. For instance: the very first transformation scene in the work of the Roman poet Ovid is also the grisliest, and one of the earliest fictional accounts of lycanthropy. It was always the story in the *Metamorphoses* that I, as an unpleasant child, loved most. King Lycaon murders a hostage boy, "cooked his limbs, still warm with life, boiling some and roasting others over the fire," and serves him to Zeus. This kind of revenge-cuisine happened so often in myths you would think the gods would have become wary, but Zeus was not. On discovering what he has eaten, Zeus strikes Lycaon's palace with lightning and banishes him into the wilderness. "There he uttered howling noises, and his attempts to speak were in vain. His clothes changed into bristling hairs, his arms to legs, and he became a wolf. His own savage nature showed in his rabid jaws." Transformation, for Ovid, is a kind of truth-telling, and the truth of the wolfish was their underhanded savagery. (Ovid was wildly popular in his own lifetime: which suggests that others eagerly agreed.)

I have always cherished fairy-tale wolves for the way in which they tell of the desires we attempt otherwise to keep hidden: about how large are our eyes and teeth and hungers. In one Russian fairy tale even more madly vertiginous than most, "Ivan Tsarevich and the Grey Wolf," the son of a tsar comes upon a wolf who eats his horse and suggests Ivan ride on his back to glory. The wolf later shapeshifts into a princess at a wedding: in some versions, he eats some guests. So a true fairy-tale wedding is not, like most British royal weddings, a reinforcement of a state institution through the medium of organ music and enormous dresses. A true fairy-tale wedding would be one in which secret desires leak out: one in which the aging prince, tired of waiting for his throne, turns into a wolf and eats the queen.

Perhaps it would have been better, though, if when looking for symbols for our rapaciousness, we had kept to dragons. The reputation of the wolf has been almost enough to destroy them. The black wolf was hunted to extinction in 1908; Gregory's wolf, a tawny, slender creature, *Canis lupus gregoryi*, died out in 1980. We have continued to hunt wolves long after they ceased to be a physical danger to us: now that there is very little rabies, the chances of a person being attacked by a wolf are vanishingly small. In fact, they're shy, cautious animals: they make atrocious guard dogs because, faced with strangers, their first impulse is to run and hide. A wolf can eat twenty-two pounds of meat in a single sitting, but their preferred food is not human but elk and deer, and also melons, figs, grains, grass; in North America, they feast on blueberries, and raspberries when they can get them. So wolves are formidably hungry, yes, just not for us.

But slowly, our understanding of them is changing. Their numbers are rising across Europe; there are currently twelve thousand pacing across the continent, shifting the landscape as they go. Wolves have worked marvels before now: they were exterminated from Yellowstone Park in the 1970s, but in 1995 they were reintroduced. Within a decade, the elk population had halved, and without the browsing elk, aspen, willow, and cottonwood trees sprang up. If you want to nurture a forest, plant a wolf.

We have watched them, and made our anxious guesses, and our guesses have largely been wrong. For instance: they don't particularly care about the moon, for it's not the night sky they are howling at, but each other. Their lives are interlinked by their calls—they howl to call each other together to hunt, to warn against threat, to find each other amid storms and ice. (I have seen one howl: they rock backward, and just before they call, they look exactly like a child about to blow out the candles on a birthday cake.) The howl can travel more than sixty miles, and on a still day, they can hear your footsteps more than nine miles away across open country. They have complex, hierarchical social lives: because lower-ranking males do not mate, "alpha wolf" denotes the male wolves who become parents. (Tell it to the next man you hear boasting of his alpha status: it just means fatherhood.)

Their sophistication extends to their communication: wolves are one of the very few animals who convey information with facial expressions. Roughly, they can be translated: ears flat back and close to the head, tail between the legs: "Don Corleone, I am honored and grateful that you have invited me to your home on the wedding day of your daughter. And may their first child be a masculine child." Ears forward and tail straight out, upright dominant posture: "Leave the gun, take the cannoli." Ears low and out to the side, teeth bared and a wrinkled snout: "I'm gonna make him an offer he can't refuse."

I once met a half-tame she-wolf on the Welsh borders. She looked far less like a dog than I had imagined she would: she had humps of muscle under her skin that you don't see in dogs. It was hard to believe that she was the wild ancestor of all poodles and pugs. She had precision, and control; she could pluck a single blackberry from a bush with her teeth. She smelled not at all like a dog: instead, of dust, and blood. Her fur, thick enough to allow her to sleep comfortably at −40°F, prickled with static when I touched her. She was very literally electric. She did not want to meet my eye. Wolves are like the fairy tales they prowl through: wild, and not on anybody's side.

THE
Hedgehog

Pliny the Elder was not an easy man. He reprimanded his nephew, Pliny the Younger, for walking the streets instead of being carried, thereby wasting hours in which he could have been reading. But in 77 CE Pliny turned the focus of his attention to the hedgehog in his *Historia Naturalis,* and gave birth to one of the loveliest myths in natural history. "Hedgehogs," he wrote, "prepare food for the winter. They fix fallen apples on their spines by rolling on them and, with an extra one in their mouth, carry them to hollow trees." St. Isidore of Seville picked up on the idea, insisting that hedgehogs collected grapes on their spines in order to carry them to their young. Charles Darwin wrote in 1867 that he had it on good authority that the hedgehog could be seen in the Spanish mountains "trotting along with at least a dozen of these strawberries sticking on its spines . . . carrying the fruit to their holes to eat in quiet and security."

Of all the wild untruths in the world that we might wish were fact, this is high on this list. In fact, hedgehogs do not eat fruit, preferring beetles, worms, eggs, and small carrion, nor do they hoard food for the winter, nor have there been any recorded instances of their using their spines as cocktail sticks. But they are nonetheless remarkable, for their place in history, for their survival skills, and for their delicate, erudite-looking beauty. Each hedgehog has around six thousand hollow spines, nut-brown at the base, rising to a strip of black and changing at the very tip to the purest white. When threatened, they roll into an impenetrable ball, which deters almost all animals except badgers, and us: Pliny wrote that you could unroll them by sprinkling boiling water on them, which does, unlike his dietary notes, seem to be true.

Aristotle suggested that hedgehogs mate upright on their hind legs, belly to belly, to avoid each other's spines. In fact, they mate like other four-legged mammals, although the process is more stressful than for most: it's not unusual for the female to move off, mid-mate, leaving the male struggling to keep up on two hind legs, slipping slowly down the rake of the female's back. Hence the old joke: How do hedgehogs mate? With great and scrupulous care. When the hoglets come, after thirty-two days' gestation, they are born utterly soft and reddish-pink, with a thin layer of skin over their spines; the spikes start to break through it immediately after birth. Within ten days they have learned how to roll into a ball; within fourteen, their eyes open, and they take on that distinctive look of polite, gracious inquisitiveness.

Although hedgehogs are ancient—they existed almost unchanged fifteen million years ago, back when we were still great apes—the name *hedgehog* is a recent invention. In Middle English they were *irchouns*—from the Latin *ericius*, a spiky military rod used for defense—or urchins. In one medieval recipe book, the "hirchone" is a kind of canapé—ground pork mixed with saffron and stuck with slivers of almonds for the spines (surprisingly reminiscent of the 1970s). When Prospero threatens Caliban in *The Tempest*—"urchins shall, for that vast of night that they may work, all exercise on thee"—he is imagining a fleet of hedgehogs, going to war on Caliban's body. The sea urchin, then, takes its name directly from the hedgehog.

We have turned throughout history to the hedgehog: we have used them in our fables, and demanded that they cure us of our pains. In 1693 the physician William Salmon published a cure for baldness, suggesting the fat of a hedgehog be mixed with that of a bear and applied to the scalp. Failing that, he suggested optimistically that hedgehog dung might have a similar effect. He was not the first to think it: the Ebers Papyrus, dating from around 1550 BCE, suggested that an amulet in the shape of the hedgehog would stop hair thinning. Its skin and spines have been thought to help, over the last two thousand years, with toothache, kidney stones, diarrhea, vomiting, fever, deafness, UTIs, leprosy, elephantiasis, and, frequently, impotence. In Latvian folklore, the hedgehog is a symbol of regeneration and fertility; Latvian wedding songs dub the bride "she-hedgehog," and married women "mothers of hedgehogs."

We have not consumed them only as medicine. In Roma families the hedgehog would be encased in clay, roasted on a fire, and, once cool, the clay would be broken open, taking all or most of the spines with it; a dish known as hotchy-witchy. In 1393, the *Ménagier de Paris* suggested that the "hedgehog should have its throat cut, be singed and gutted, then trussed like a pullet, then pressed in a towel until very dry; and then roast it and eat with cameline sauce"—a sauce including bread, wine, vinegar, cinnamon, and ginger—"or in pastry with wild duck sauce." I have a friend who has caught and eaten hedgehog: it tastes, she says, like anorexic rabbit.

Even today, when they are a protected species across most of the world, we struggle to let them alone. In the United States, though there are no living species native to the continent (at least, not for the last five million years: the extinct *Amphechinus* ambled through North America during the Miocene period), they're imported as cute pets. (Though not if you live in the five boroughs of New York City, where hedgehog husbandry is illegal.) There are cafés in Tokyo where you can dress a hedgehog in a hat and handbag for a photograph. I have been to one, once: clad in their felt bonnets, although carefully treated by their handlers, the hedgehogs do not look as though they find the situation ideal.

In England, where they go without hats and accessories, their numbers are falling, and have been doing so for decades. The loss of hedgerows, vast open fields without cover, and death by cars are in part to blame, alongside the mass use of pesticides and global warming reducing their insect prey. Despite pledges from tens of thousands of people to make their gardens more hedgehog friendly (by leaving a small gap in every fence, no larger than a saucer, to allow them to roam), there are currently perhaps a little under a million left— a ninety-seven percent drop from the thirty million that roved the UK in the 1950s. Thirty million is far more than the current number of pigeons in Britain; when my parents were young, hedgehogs were everywhere, a proliferation of commonplace spinose beauty.

In 2015 the Tory MP Rory Stewart gave an impassioned thirteen-minute speech to a largely empty House of Commons, entirely about hedgehogs. It was the first time the hedgehog had been discussed by Parliament, he said, since 1566. (In fact, Stewart wasn't quite right: in the 1650s, Sir Richard Onslow leveled an attack on King Charles I's foreign policy, saying he had like a hedgehog "wrapped himself in his own bristles," but that was only in passing.) That 1566 debate led to the decision to put a tuppence bounty on hedgehogs: farmers believed they suckled milk from their cows in the night, and as a result up to two million were hunted and handed in.

That was another of our mistakes: hedgehogs are lactose intolerant, and milk can kill them. In fact, they prefer insects and water: they were kept in some Victorian kitchens as a form of pest control, to keep down the cockroaches. They're a peculiar mix of tough and delicate: they are immune to most snake venom, but occasionally suffer from a condition known as "balloon syndrome." A glottis at the top of the windpipe, which opens and closes, can become stuck, leaving the air nowhere to go; the hedgehog inflates to more than twice its usual size, and has to be punctured like a balloon. Come the autumn, there is the added risk of hedgehogs taking up residence in bonfires and, on the Fifth of November, being burned alive.

If we were not used to hedgehogs—if they existed only in Yosemite or the Okavango Delta—we would surely travel thousands of miles to see them, such is their peculiar loveliness. These are hard times, and the world is already aflame. The least we can do is refrain from setting alight some of the world's sharpest and gentlest creatures.

THE
Elephant

I n 1870, the Prussian army laid siege to Paris. Its defenses were formidable, so rather than fighting, the Prussians, led by Wilhelm I, chose to ring the city round with a blockade and starve its people into submission. The hunger made Parisians both desperate and inventive. A rat, smoked and dressed with spices, could fetch two francs, while a cat might be worth twelve. A luxury grocer, owner of the Boucherie Anglaise on the Boulevard Haussmann, approached the zoo, his eye on the two male elephants. A deal was struck: for twenty-seven thousand francs, Pollux and Castor were sold. Because nobody had experience of slaughtering elephants, a marksman was hired to shoot them with steel-tipped explosive bullets. They were skinned and sold at staggering prices to Paris's richest citizens. Henry Labouchère, an English politician and theater owner, and author of *Diary of the Besieged Resident in Paris*, wrote, "Yesterday I had a slice of Pollux for dinner. Pollux and his brother Castor are two elephants that have been killed. It was tough, coarse, and oily. I do not recommend that English families eat elephant, as long as they can get beef or mutton."

The trunks of Pollux and Castor, the tenderest part, were sold as a delicacy for forty francs a pound. The famished citizens who ate them were, unknowing, consuming a marvel. An elephant's trunk is a fusion of the upper lip and nose, freighted with 40,000 muscles (we have about 650 in our entire bodies). What looks like it must be an unwieldy, erratic kind of appendage is, in fact, under the elephant's calm and confident control; with its prehensile tip, the African elephant can pluck a single blade of grass, or lift 350 kilos, or swing a man into the air. Their two thousand olfactory receptors (bloodhounds have a paltry eight hundred) mean they can smell water two miles away; a small group of African elephants have been trained to sniff out land mines in Angola. Faced with deep rivers, elephants can use their trunks to snorkel, wading and swimming underwater with the tip held carefully aloft. And it's the trunk that allows them to give their great wild trumpetings, when scared or aroused or spoiling for a fight.

It's not with the trunk alone, though, that elephants make their most remarkable sounds. Elephants are among the very rare group of living things—whales being the most famous— who can communicate using sounds of such low frequency that they're pitched below the range of human hearing. The lower a sound wave's frequency, the farther it can travel; with their calls, known as infrasounds, elephants use their immense larynxes to produce a note so low that they can be heard by others six miles away. They can also produce infrasound waves that pass into the ground, which can be felt in the feet by herds dozens of miles away—which allows elephants in their mating season to locate each other over huge distances. What looks to us like a pensively silent elephant may be exchanging information about predators, or water resources, or sex—an evolved telegraph system in the wilderness.

The elephant's trunk is also its vulnerability. We have believed, on and off since about 77 CE, that elephants are afraid of mice, because the mouse might climb up its trunk and settle in. John Donne wrote a poem in 1601 that described a malicious mouse climbing up the trunk of an elephant and eating the brain:

> Nature's great masterpiece, an elephant
> (The only harmless great thing), the giant
> Of beasts . . .
> His sinewy proboscis did remissly lie,
> In which, as in a gallery, this mouse
> Walked and surveyed the rooms of this vast house,
> And to the brain, the soul's bedchamber, went,
> And gnawed the life-cords there. Like a whole town
> Clean undermined, the slain beast tumbled down.

In reality, elephants are not at all afraid of mice, unless startled by them in poor light—but they are terrified of bees. Bees fly inside their trunks and sting the soft tissue there. Elephants have been seen fleeing from swarms, ears flapping, trumpeting in fury and pain. Their fear has recently led to a successful strategy in some southern African countries, in areas where elephants are a nuisance to crops and elephant-human conflict is rife, to establish beehives around the outskirts of fields: the bees act as sentinels, the elephants go foraging elsewhere, and the villages are provided with plentiful honey.

Pliny—not known for his sentimentality—believed they were among nature's sweetest creations. "The animal's natural gentleness towards those not so strong as itself is so great that if it gets among a flock of sheep it will remove with its trunk those that come in its way, so as not unwittingly to crush one." Pliny's theory had some truth in it: their gentleness, and their sense of care. Elephants returning to a matriarchal group are greeted with ceremonial embraces, entwinings of trunks and colossal snorts of pleasure. Coming across the bones of their dead, they will salute them, lightly touching the skulls and tusks with their trunks and vast, heavy feet. They have been known to bury dead members of their herd, covering them in earth and leaves.

There is nothing on the land as large as a large elephant—they are the greatest of us. The biggest ever recorded was an African male in Angola, standing thirteen feet at the shoulder, and weighing twenty-four thousand pounds (the weight of a garbage truck carrying a vending machine and a grand piano). The Asian elephant, with its dented two-domed forehead and more tractable manner, is smaller, but still of prodigious size; an average male would stand at nine feet, eye to eye with Robert Wadlow, the tallest man in the world. And then there is a subspecies of the Asian elephant, the Borneo elephant, known locally as the pygmy elephant. Living only in the northern parts of the island of Borneo, it's not in fact so very tiny—about thirty percent smaller than the mainland Asian—but its round face, short trunk, and disproportionately large ears give it a scampering, miniature look. Its tail is so long that it sometimes scrapes along the ground, leaving a wobbling line in the dust as they go, like a very long arrow pointing after them: Elephants This Way.

Nobody is quite sure from where, exactly, this smallest of the elephants came. For many years the theory was they had been introduced to Borneo in the eighteenth century by the Sultan of Sulu, and were descendants of a unique, domesticated breed of elephant, selectively bred for their petite size. They are much gentler than their mainland counterparts, and more curious and investigative, which was offered up as evidence of their having once lived in harmony with man. They have, though, no love of human intervention in their forests: traps set for smaller animals by local hunters, even if they pose no threat to the pachyderm, will be carefully and thoroughly trampled by any elephant who finds them.

More recently, genetic analysis has discovered that the Borneo elephant probably arrived on the island long before the sultan; it's likely that during the Pleistocene epoch, when the ice age created a bridge between mainland Asia and Borneo, the elephants processed across. When the ice melted, they held sway. For three hundred thousand years, the Borneo elephant has been the largest mammal on the islands—albeit on the smaller side of vast. They have adapted to Borneo's heat, plastering themselves and each other with mud as sunscreen during the summer months. Latterly, though, deforestation for palm oil plantations has laid waste to much of their habitat, and they find themselves edging closer and closer to humans, trampling crops and destroying livelihoods. Hundreds of humans and elephants die each year as we meet in too-close encounters: we are not suited, as things stand, to be good neighbors. There are, at an optimistic estimate, 1,500 Borneo elephants left. Poaching for tusk, skin, hair, and meat, and our constant movement into green uncultivated space, render their return in numbers almost impossible. Where there was once a swath of beauty, there will be thin scatterings of it. They are one victim of a far larger conundrum: that we have not yet risen, as a human species, to the concept of that which we cannot undo.

The elephant-headed deity Lord Ganesha is, in Hinduism, the most sacred Remover of Obstacles: a glorious figure, he is the god of beginnings. In one version of his story, he was made by the great goddess Parvati; in order to prevent anyone from disturbing her while she was bathing, she formed a child from turmeric paste and the dry skin she scrubbed from her arms as she washed. She set him to guard outside her room, and so when the god Shiva, her husband, tried to enter, Ganesha obediently barred the way. But Shiva, angered, struck off his head. Parvati, heartbroken and furious, ordered Shiva to restore her child to life. Repentant, he sent out his men to bring the head of the first creature they saw, which happened to be an elephant. Ganesha is, too, the *deva* of wisdom: it was he who formed a pact with the sage Vyasa to transcribe the great epic Mahabharata. In order that it not take them forever, it was made a condition that Vyasa would dictate and Ganesha write simultaneously, and neither would stop. After three years of continuous writing and speaking, the epic was nearly complete, when Ganesha's feather pen broke—and so without pausing he snapped off his tusk, dipped it in ink, and continued to write. He is, therefore, a patron of the arts and sciences. Portrayed with one broken tusk, the elephant-headed god salutes learning, and the hope it brings—two unwieldy qualities of which we will have an exponentially increasing need in our years immediately to come.

THE

Seahorse

P oseidon knew how to travel. His underwater chariot was flanked by nymphs—up to thirty-three, if you believe Homer, or fifty, if you believe Hesiod—and pulled by a team of seahorses. Virgil wrote of Neptune's seahorses:

> *Where'er he guides*
> *His finny coursers and in triumph rides*
> *The waves unruffle and the sea subsides.*

The poet Robert Herrick, author of many breezy lyrics about death, imagined the seahorse as the ideal heroic steed:

> *Give me that man, that dares bestride*
> *The active sea-horse, and with pride,*
> *Through that huge field of waters ride,*
> *who, with his looks too, can appease*
> *The ruffling winds and raging seas.*

It was said that ancient Greek fishermen, untangling seahorses from their nets, believed they had in their hands the newborn young of Poseidon's steeds. The largest in the world, the big-belly seahorse, would in fact at a foot long be just large enough for a human baby to ride, if both were so minded. The smallest, Satomi's pygmy seahorse, would at about half an inch not cover the top joint of your thumb. But any seahorse, no matter the size, would be more than fit conveyance for a god—because gods crave our awe, and everything about the seahorse is a stark astonishment.

The seahorse is the only species in the animal kingdom in which the male gives birth. The female deposits her eggs in the male's abdominal pouch, in a process that looks like a more intimate version of using a postbox. He fertilizes them as they enter and keeps them safely gestating for between two and six weeks. The process of giving birth is both triumphant and disconcerting. It's more akin to a confetti cannon than most other births; the male seahorse appears to convulse, as if sneezing or vomiting, and from the opening at the top of his womb erupts a herd of minute seahorses, up to 1,500 fry, until he disappears in a cloud of his own offspring. Less than 0.5 percent of the tiny young will survive into adulthood, which is probably why the male takes on gestation: it allows the female to immediately begin making another batch of eggs, which allows for more pregnancies during the breeding season, which allows for more fry, and more possibilities of life.

Many seahorses pair once and forever. Finding a seahorse mate, amid waves and swirling vegetation, is hard; their superb camouflage means they are hidden even from each other, and that, coupled with their inability to travel with speed or precision through the ever-moving sea, makes finding a partner exceptionally stressful; by remaining faithful, they gain time in which to undergo more pregnancies and increase their chances of reproductive success. This is not perhaps as romantic as it might be—but they also dance. Each morning, the two meet in the male seahorse's territory: as they near each other, their colors change, out of subtle camouflage shades and into vivid hues: brown into white, white into yellow. The seahorse has tiny cells, chromatophores, embedded in the skin, each containing liquid pigment. By contracting or expanding the chromatophores, different colors appear with different intensity: orange, pink, red—a little like playing a color organ. They circle each other, newly bright; the male twists around the female; their tails entwine. As they move, they click at each other (one of two noises of which they're capable: the other is a minute guttural growl when threatened, so faint as to be almost inaudible to human ears). Then the female returns to her own territory, until the next day, when they dance again. For writers, hermit types, and part-time misanthropes, this might sound like an ideal marriage.

A seahorse is technically a fish, but then so is a shark, so that doesn't tell you much. Unlike sharks, they are perilously fragile. Easily buffeted by the sea, they have a single fin on their backs to propel them forward; the two small pectoral fins up behind their eyes are for steering only. The back fin beats to and fro up to fifty times a second, but even so progress is slow and exhausting: akin to standing on roller skates and attempting to propel yourself forward by waving a copy of the UN's terrifying *Global Assessment Report on Biodiversity and Ecosystem Services*. In storms, they can be tossed and spun and exhausted to death. Other factors also militate against their easy passage through life: having no stomach, they must eat almost constantly to stay alive.

They belong to the family Syngnathidae, from the Greek *syn*, together, and *gnathos*, jaw: their fused jaws mean they cannot chew; instead, they hoover plankton and small crustaceans through their long snouts. It's thought that the seahorse's shape might be explained by tectonic shifts thousands of years ago, when the Earth's movement created shallow waters in which "meadows" of sea grasses could thrive. The seahorse might be a highly evolved version of the unprepossessingly worm-like pipefish; the grasses, growing vertically toward the sun, may over time have caused a pipefish-like swimmer to shift to an upright posture as it moved through the sea-meadow. And for all their fragility, they're exquisitely evolved: the strength and flexibility of their tails mean they can latch themselves to coral or to the roots of grasses to rest, or hitch themselves to patches of floating vegetation, and go traveling through the ocean at what is, for a seahorse, breakneck speed.

They look mythic. The leafy seadragon, which looks exactly as impossible as it sounds, can transform to match the seaweed among which it hides; the Pacific seahorse can move from gold to maroon. This ability, though, has made the task of determining how many species there actually are a difficult one: for a while, some researchers thought there might be two hundred species in the world, while others thought it might be only a few dozen. We think there are forty-seven, but we risk discovering them only to lose them; twelve are vulnerable, seventeen are listed as data deficient, and two as endangered. Numbers are falling over the world; the population of seahorses in the Philippines has dropped by nearly three-quarters in ten years. They're caught up in trawling nets meant for other fish—nets which also devastate the ocean floor, tearing up their territories—and are either thrown away or dried and sold on to China and Taiwan, where twenty million are consumed every year, and where medicinal use of seahorses dates back two thousand years. Until urgently needed legislation outlaws these forms of trawling, we will need to refuse to eat anything that is taken from the ocean by over-exploitive non-selective fishing. More invisible, and more deadly, our rising sea temperatures mean they may not have time to move to cooler waters, leading to mass deaths. The possibility that the majority of species could, by 2050, be in fact mythic is very real.

We live in a world of such marvels. We should wake in the morning and as we put on our trousers, we should remember the seahorse, and we should scream with awe and not stop screaming until we fall asleep, and the same the next day, and the next. Each single seahorse contains enough wonder to knock the whole of humanity off its feet, if we would but pay attention.

THE

Pangolin

I did not believe in love at first sight, but I have found there's an exception, and that is the pangolin. To reach this particular pangolin was difficult, which felt only reasonable; something so remarkable shouldn't be gained with ease. She lives in a wildlife conservation project outside Harare, Zimbabwe. The roads in Harare have been deteriorating for years; gaps are patched with house bricks, and during the rains it would be possible to bathe a Great Dane in the potholes. Most of the road signs have been long since stolen—the rumor was that they were used for coffin handles during the cholera outbreak in 2008, although that's probably mostly apocryphal—and you drive by guesswork and hope. Hot pink and purple bougainvillea grow at the verge, and occasionally overwhelm the traffic lights.

The pangolin is known as a scaly anteater, because of its diet, and because it's the only mammal entirely covered in scales, but the description does not acknowledge the fact that the scales are the same shade of gray-green as the sea in winter, and the face that of an unusually polite academic. The tongue of a pangolin is longer than its torso, and it keeps it tidily furled in an interior pouch near its hip. The name comes from the Malay word *penggulung*, meaning "roller"; when threatened, they curl into a near-impenetrable ball. Their defense mechanism has made them easy prey to humans—rather than offering protection, it renders them neatly and readily portable.

The Sangu peoples of southwestern Tanzania heralded the

arrival of a pangolin as a great event; it was said in the folklore that they fall from the sky, sent by the ancestors, and follow home the first person they meet in the bush. If a pangolin did arrive at a person's home, it was treated with both respect and careful wariness: with songs, a slaughtered sheep, the pangolin clad in a black gown, and a ceremony that ended in dancing. It was said that the pangolin had been known to dance, too, rising on its hind legs. In most accounts of the ceremony, which date largely from the 1950s, the pangolin was slaughtered, wrapped in the black cloth, and buried: returned to the ancestors. In Zimbabwe, folklore initially militated against the killing of pangolins. Pangolins were harbingers of the purest and finest kind of luck, and mothers still tell their children that pangolins are the source of alluvial gold sometimes found in the soil: pangolin droppings, digested ants turned treasure trove.

Latterly, though, rural communities in Zimbabwe, where 1.6 million children live in extreme poverty, have been offered an alternative source of income, in the form of pangolin hunting. The highways in Harare have for years been lined with billboards blazoned with images of pangolins and the government-sponsored motto: "Wildlife trafficking is a crime. Africa is an adventure—enjoy it, don't destroy it." The billboard's ask is a hard one: the incentive is so high, and the need is so great, and pangolins are currently the most trafficked animals in the world. Their scales are used in traditional Chinese medicines and their flesh is eaten as a delicacy; roasted pangolin meat is thought to stimulate lactation and improve blood circulation. A 2007 report in *The Guardian* quoted a chef in Guangdong on the means of preparation:

> *We keep them alive in cages until the customer makes an order. Then we hammer them unconscious, cut their throats and drain their blood. It is a slow death. We then boil them to remove the scales. We cut the meat into small pieces and use it to make a number of dishes, including braised meat and soup. Usually the customers take the blood home with them afterwards.*

Of the eight species of pangolin, two are listed as critically endangered in the International Union for Conservation of Nature's Red List of Threatened Species. Guangdong customs have seized as many as seven tons of scales at a time being shipped into China; each ton will have required the death of 1,660 animals. It is a fact so exhausting, so dreary, that it's difficult to fathom.

For now, though, they continue to forage and roam and mate—and, in some cases, to climb. The long claws and long prehensile tail of the African tree pangolin—one of the eight extant species—allow it to walk straight up branchless tree trunks with the ease and casualness of a Sunday stroll; for all that they are low-slung and short-legged, they are unintimidated by gravity. As with all species, the newborn tree pangolin young—known as a pup, or to the fancifully inclined, pangopup—rides on its mother's back for the first months, rising to great arboreal heights as it clings on near the base of her tail. To sleep, the baby will often roll into a ball, and the mother will form her ball around him: pangolin matryoshkas.

This particular pangolin, in Zimbabwe, has a keeper—a man who walks with her through the bush from anthill to termite mound for ten hours a day, keeping her always in sight. She needs to consume roughly seventy million insects a year in order to stay alive. If the distance is relatively short, she walks. If it's longer, she is carried in her keeper's arms, or in a specially designed backpack. He set her down on the ground to demonstrate how a pangolin walks; she moves on her hind legs only, her forelegs raised and her long claws clasped together in front of her, as if knitting her fingers together in anxious thought. She returned to her handler, setting one hind foot on his shoe in order to allow herself to be more easily lifted onto his shoulder. It was unlike anything I have ever seen. Her loveliness makes other forms of loveliness—diamonds, rubies, wrists bedecked with Rolexes—look like a con.

Pangolins are more beautiful than seems plausible in this fallen old world; they look as though they should be strictly prelapsarian. And they are in fact prehistoric; they are eighty million years old, in contrast to our six million. They were here first, at the origins of things; they have an ancient, unassailable, ungainsayable right to remain.

THE
Stork

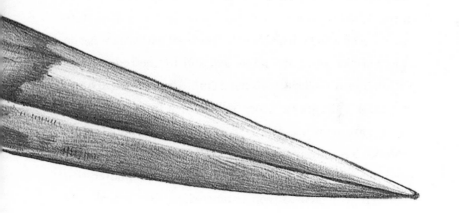

I t was wartime, and propaganda fell from the sky like rain. Nazi planes dropped leaflets over British lines in Europe, telling the men that their wives were in bed with American soldiers, complete with drawings of said wives undressed. The Allied forces flew hydrogen balloons over Axis troops to scatter images of fields lined with German graves. But the scope of both planes and balloons was limited. So when Himmler wanted to send propaganda to the Transvaal in a bid to win the support of the Boers, he ordered his scientists to investigate the possibility of using migrating storks as carriers. Test flights for the *Storchbein-Propaganda* began, until it was found that a thousand birds would be required for every ten leaflets that reached their target. The plan was abandoned. Others, on the Allied side, were more persistent. In 1940 a dead stork was found on a farm in the North Transvaal with a message on a piece of tape sewn around its leg, sent by the resistance in Nazi-occupied Holland: "To our South African brothers: we, the people of Bergen op Zoom, tell you that living under German occupation is just hell."

Storks have always been our carrier birds, and we have loved them for it—though exactly why we told children that they were delivered to their parents slung from a beak isn't wholly clear. The story may have originated in Slavic mythology, where the stork carries unborn souls from Vyraj, a spring paradise, to earth. It's also possible that mistaken identity is involved: in Greek myth, Zeus's consort Hera transforms the Pygmean queen Gerana into a leggy long-beaked bird; Gerana then rescues her baby from Hera's clutches, carrying it in her beak. But the bird in the original myth is a crane (*geranos*), not a stork. The version we know best comes from Hans Christian Andersen. His rendering is more heavy metal than ours: a group of young storks are taunted by a cruel child, so their stork mother tells them she knows the pond "in which all the little babies lie, waiting till the storks come . . . We will fetch a little baby brother or sister for each of the children who did not sing that naughty song." But for the cruel child "there lies in the pond a little dead baby who has dreamed itself to death. We will take it to the naughty boy, and he will cry because we have brought him a little dead brother." This bit of the story is generally left off greeting cards.

In 1822 storks solved the mystery of where birds disappeared to in winter. The question had puzzled ornithologists since the time of the ancients: Aristotle had been pretty sure that storks hibernated in trees. He also deduced that redstarts transformed into robins in the winter months and turned back again in the spring. In this he was no more wrong than Olaus Magnus, Archbishop of Uppsala, who in 1555 reported that swallows slept out the winter at the bottoms of muddy lakes. Indigenous North American accounts told of hummingbirds hitchhiking on the backs of geese; Homer suggested that every spring cranes went to war against "the pygmy-men" at the ends of the earth—revenge, he said, for Hera's mistreatment of their queen. In 1694 the nonconformist minister, scientist, and vice-president of Harvard Charles Morton suggested in deadly earnest that the stork, along with the swallow and crane, wintered on the moon. Then, in 1822, a stork arrived in a German village with a thirty-inch spear in its neck. The spear, metal-tipped, rising up through the bird's breast and out through the side of its neck, was identified as coming from central Africa. The arrow stork, *Pfeilstorch*, was the proof we had been waiting for: birds were flying halfway around the world every year, returning in the spring. (Far more unlikely and fairy-tale-like, really, than roosting inside a local tree.)

Many of our stories about storks are love stories, extravagant salutings of their intelligence and heroism. There is a great deal to love: they're big, the Hercules of birds; of the nineteen species, the largest, the African marabou stork, can reach five feet tall, with a wingspan of ten feet. They look wise: the white stork has a flick of black eyeliner that gives it a look of knowing intellect. In 1536 the city of Delft was half consumed by fire; a Dutch physician, Adriaen de Jonghe, saw a female stork return from hunting to find her nest in flames. She tried to lift her babies out of the nest: failing, she covered them with her body and allowed herself to be burned along with the young she was powerless to save. In 1820, it was reported that storks extinguished the flames that ran through the town of Kelbra—though the alleged author of this claim, the "little known" and possibly apocryphal Okarius de Rudolstadt, doesn't say how that might have worked. Storks appear even at the Crucifixion. They're largely mute, but in Scandinavian lore a stork is said to have wheeled and circled above the cross, crying out in one great effort *"styrka, styrka!"* ("strengthen ye!" in Swedish). Hence, apparently, their name: the same in Swedish and English.

America's only native stork, the wood stork, which has tail feathers of exquisite purple-green and the crepuscular neck of a thousand-year-old heavy smoker, seemed in the 1980s to be destined for extinction. Its numbers spiraled downward from 150,000 to 10,000 after the habitat of three-quarters of America's storks, Florida's Everglades wetlands, was eroded by human development. But the storks, with the help of activists, were taught to move north, and make new homes, in wetlands, woodlands, and, in one case, just off an airport runway. They can now be found across the country: in Alabama and Mississippi, in Georgia, South Carolina, and North Carolina. They're not wholly safe—they are listed in North America as threatened—but they are a true success story of back-from-the-brink.

Our admiration for storks has swung—haphazardly, destructively—between the sentimental and the gastronomic. Until very recently, Britain was a country without any storks at all. The penultimate English-born stork hatchlings were in 1416; then there was a 604-year wait until May 2020, when five chicks were born to one of the hundred or so birds introduced as part of a rewilding project on the Knepp Estate near Horsham in Sussex. Nobody knows why they became extinct in Britain: it's said they prefer republics, so we could blame the royal family, but it's more likely they were hunted into nothing for food. They featured in medieval feasts as one of the ingredients of game pie, a delicacy which could also include heron, crane, crow, cormorant, and bittern. In Europe, they were part of the ritual of spectacular dining well into the seventeenth century: food gilded with precious metal, cocks wearing paper hats mounted on pigs' backs like jockeys, boars' heads with fireworks shooting from their mouths, and storks roasted and then replumed to look as if they had just folded their wings from flight and come to rest on the table.

That flight—effortless, barely flapping—could be credited with bringing us human aviation, since the great nineteenth-century aeronaut Otto Lilienthal built his experimental gliders based on the movements of storks. He studied the way their wings moved, how easily they soared on thermals, how they took off into the wind, the way their wings tapered to a point and were exquisitely cambered in cross-section. "The impression could be given," Lilienthal wrote, "that the only reason for the creation of the stork was to awake in us the desire to fly, to act as a teacher to us in this art." Leaping off the Rhinow Hills in his stork-like plane in 1893, he was able to travel 820 feet: enough truly to know flight. He died three years later when his glider stalled in mid-air, an accident for which the storks are indirectly responsible.

Storks are said to bring luck to the houses on which they roost. (They're also a fire hazard: they build nests up to six feet across and ten feet deep, returning year after year to add to them.) I once saw one in Zimbabwe kicking up insects from the grass and catching them in her long beak—a sight to blow your hair back. Standing, all legs and beak, they look like a stroke of calligraphy by a flamboyant and ambitious god. Even the ugly ones—like the marabou, which with its tufts of hair and big neck pouch hanging like a scarf or a testicle has the aspect of a disreputable undertaker—are beautiful. They produce marvels without warning: when the woolly-necked stork opens its wings in flight, it reveals a band of unfeathered skin on the underside of the forearm that shines a startling ruby red. Clattering life-affirmers, hope-birds; it's said in Eastern Europe that the excitable rattling of their bills is applause for the oncoming summer. Their wings in the spring sky read as semaphore, beating through the air, proclaiming: "Strengthen ye, strengthen ye!"

THE
Spider

I like spiders and I like gymnasts, but I do not like them both at once. Be one or the other but not two together, I'd ask. And yet the jumping spider exists, and it is magnificent, in its way. We have not been taught to love things with four pairs of eyes, but there is greatness in them.

Of all the forty-five thousand species of spider, the jumping spider is perhaps the most fiercely brave: where black widows prefer to hide from humans, jumping spiders will advance and investigate. A jumping spider the size of my little fingernail can jump upon and kill a large grasshopper, which is roughly equivalent to my leaping upon and devouring a Volvo station wagon. And they are many; the jumping spider clan is the largest in the spider world. Belonging to the family Salticidae, there are 610 genera and more than 5,800 described species, about thirteen percent of all spiders. They are the tigers of the arachnid world, ferocious and nimble; some can jump up to forty times the length of their own bodies. As with all spiders, and in contradistinction to our vulnerably fleshy example, the salticid keeps its muscle safe inside its bones, the exoskeleton enveloping and protecting the muscle. The leg muscle, though, is not particularly strong. Instead, the jumping spider is a miniature hydraulic pump; it contracts the abdomen, forcing bodily fluids, largely blood, into the back legs, which makes them straighten, catapulting the spider forward. As they leap, they tether a dragline of silk to their jumping-off point; if the jump fails and the prey escapes, they can winch themselves back up to safety, unhurt, unembarrassed. Their blood, as befits their status, is blue.

I understand that spiders are hard to cherish. They appear to be all right angles and stubble, and they do not sleep, and their eyes never close, and we find it difficult to admire that which does not blink. Between three and five percent of the global human population is arachnophobic, although the propensity isn't evenly spread: it's much less common in tropical places, where large, hairy spiders abound. It used to be thought that arachnophobes were particularly sensitized to the movement cues in spiders' legs: no evidence was found for that, nor was fear of spiders associated with an increased predisposition to fear in general. It does seem to run in families, and there may have been some small evolutionary advantage in the increased wariness it brought (although the panic accompanying the fear wouldn't have been helpful): but why some are phobic and others not remains, largely, a mystery. But—if you brace yourself in the right way, the spider family includes specimens that are superbly beautiful. The male coastal peacock spider, for instance, dances to impress the female: it raises its radiantly colored abdomen, marked in red and blue, revealing stark orange hairs along the edge of the back part of the body, which are visible at no other time in its life. One jumping spider, *Jotus karllagerfeldi*, is so called because its exquisite black-and-white markings—black eyes, black-and-white pedipalps under the jaw—were so reminiscent of the designer Karl Lagerfeld, in his sunglasses and white collar.

The jumping spider is an amateur, though, when it comes to production of one of the most remarkable substances on the planet: at silk making, the master is the golden orb spider, of the family Araneidae. The golden orb spider weaves yellow webs which shine like precious metal in the sun and, once cast, can stand for several years: the webs are strong enough to occasionally entangle birds mid-flight. Early fishermen in New Guinea were said to weave their silk into nets, tough enough to haul dozens of fish in one swoop. Scientists are currently trying, thus far not quite successfully, to mirror its structure for bulletproof vests.

Spider's silk is a miracle thing; one we have long tried to replicate and cannot, proof that our invention, daring and beautiful and miraculous as it is, is no rival to that with which the world already thrums. Spider silk weighs almost nothing— a thread of silk long enough to loop the Earth would weigh a pound or less—but is one of the strongest materials on the planet: five times stronger than a strand of steel of the same thickness. The silk that comes from any spider's spinneret is liquid, but becomes solid on contact with air, and is so consistently fine that it was used in the Second World War to make the crosshairs in military gunsights. If we were to make a vast web out of spiders' silk as thick as a ballpoint pen, it would halt a Boeing 747 mid-air.

It is important, though, to make sure you have the right

spider silk for the right task: in 1709, François Xavier Bon de Saint Hilaire, a minor French court figure, gave Louis XIV the first known pair of spider-silk stockings of unknown species, made from the silk of hundreds of the cocoons in which a female spider lays her eggs. They were silver-gray and shimmered in the sunlight; but once the Sun King put them on and tried to walk, they fell to pieces between his royal toes. Also, François found that the propensity of the spiders once collected in cages to eat each other made the whole enterprise a trying one.

Others have been more lucky. In 1863 the American Civil War surgeon Burt Green Wilder, coming upon a golden orb spider's vast nest, trapped it in his hat and ran with it back to his tent. There, he wrote, the spider dropped from his wrist on a length of its silk.

> *Rather than seize the insect itself, I caught the thread and pulled. The spider was not moved, but the line readily drew out, and, being wound upon my hands, seemed so strong that I attached the end to a little quill, and, having placed the spider upon the side of the tent, lay down on my couch and turned the quill between my fingers.*

After an hour and a half, he had collected 150 yards of "the most brilliant and beautiful golden silk I had ever seen." He devised a tiny set of wooden stocks, in which he could trap spiders in order to milk them of their silk; his dream of making a gown for his lady love was abandoned only when he calculated that it would take five thousand spiders. More manageably, Lieutenant Sigourney Wales, a soldier in the same regiment as Wilder, collected the same golden silk and wound it into coils. These, he claimed, he was able to sell off as gold jewelry: spider-made wedding rings.

And there is this: without spiders, we would have global famine. They eat that which would eat our food; they stave off pestilence. Spiders feed on an estimated four hundred to eight hundred million tons of insects and other pests each year; in comparison, humanity consumes about four hundred million tons in meat and fish. They eat more insects than birds and bats combined: they also pollinate plants, recycle dead animals and vegetation back into the earth, and, in turn, provide the main diet of three thousand to five thousand bird species. Without spiders, we would perish. We should salute them with gratitude: which is not hard, because, for now at least, they are everywhere. As a child in Zimbabwe I would often find spiders under my pillow, like a coin from the tooth fairy but with eyes and teeth. One field in Wales had more than a million spiders in an acre of land: the average may be closer to three million per acre in tropical climates. There are so many, legend has it, that you eat eight whole spiders a year in your sleep—a fact which would be mildly horrifying, if it were true. It was claimed by the fact-checking website Snopes that a computer columnist, Lisa Birgit Holst, wrote an article in 1993 about how fake "facts" were rapidly circulating via email chains; she offered up "eight per year" as proof of how readily false statistics can seem believable, and the spider "fact" took on a runaway life of its own. (In fact, Snopes was indulging in exactly the same tactic it ascribed to Lisa: Lisa Birgit Holst is an anagram for *this is a big troll*, and she, too, is a fake fact.) Amid all the Internet point-scoring, though, one thing is certain: spiders have no interest in your open mouth, which is moist and quivering and unwelcoming, and will not willingly go near it. There is just an inch of truth in it: you will have inhaled millions of microscopic pieces of spider in dust—but then, by that reckoning, you have inhaled substantial portions of people, too.

Even spiders, though, have their endangered members. The Gooty tarantula, which survives now only in a small cluster of forest in India's Andhra Pradesh and moves between treetops, up away from human eyes, is a staggering Yves Klein blue— nothing else in the world is that precise blue, and nothing will be, if their slide from critically endangered toward extinction continues. The tiny spruce-fir moss spider, rich yellow-brown and the size of a ball bearing, constructs tube-shaped webs in which it lives up high in the Appalachians. The population of Fraser fir trees among which they live has been decimated by logging and disease, and with them their attendant spider. The main population of the world's spruce-fir moss spiders appears to live across a single rock outcrop in North Carolina. Though so small a whole convocation could gather on your palm, their vanishing from the planet would not be a small loss: that something which once existed in the universe should be discarded, irreversibly, by our uncareful encroachments—that is large enough.

In our long history of getting a great deal wrong, we have lavished many of our errors on spiders. We have ascribed to them dangerous powers they rarely possess (less than one-tenth of one percent of all spiders have been responsible for human deaths). During the sixteenth and seventeenth centuries, it was thought that a bite from a specific kind of wolf spider (named, confusingly, *Lycosa tarantula*, after the Italian region of Taranto, but distinct from the species Tarantula) was potentially deadly: the only cure was thought to be for the victim to leap into a frenetic dance. Francesco Cancellieri, a prolific Italian writer of the eighteenth century, reported:

> *. . . we found the poor peasant oppressed with difficult breathing, and we saw that his face and hands had started to turn black. And because this illness was familiar to all present, a guitar was brought . . . First he began moving the feet, then the legs. He stood on his knees. Soon after an interval he arose swaying . . . in the space of a quarter of an hour he was leaping, nearly three hands from the ground . . . and before an hour, the black had gone from his hands and face, and he regained his natural colour.*

Special curative musical scores, such as the seventeenth-century polymath Athanasius Kircher's *Antidotum Tarantulae*, were prepared in anticipation; and it might be that the frenetic spider-jig is the origin of the stamping, heel-toe dance of the tarantella.

Recently, scientists at Manchester University trained a jumping spider called Kim to leap on their command. Her accuracy was astonishing: she never missed the target platform. Jumping spiders, with their eight eyes, can see a broader spectrum of colors than we can; the world looks different for them, more riotously technicolored. It was also found that some become fixated on nature programs, more so than by, for instance, the evening news. They are so much cleverer than we knew. It's a thing worth knowing: that so very rarely do we discover that any living creature is simpler than we thought.

Bat

I f you seek an easy recipe to become invisible, capture your-self a bat, a black hen, and a frog. Extract the hearts, tie them together, and fasten them with string under your right armpit. This, according to the nineteenth-century Parisian occultist Émile Grillot de Givry, will render you invisible to everyone but yourself. Bats have been popular invisibility ingredients for hundreds of years, on every continent. An eighteenth-century book of magic, spuriously attributed to Albertus Magnus, instructed: "Pierce the right eye of a bat, and carry it with you and you will be invisible"; in parts of Trinidad, drinking the bat's blood had the same effect. That these would all have been possible to fact-check did not put a dampener on their popularity. (But then, I've bought many creams which promise to reverse aging, with precisely identical efficacy.)

Night-fliers that they are, we associate bats with dark deeds. In Aesop, a bat borrows money for a business venture; when it fails, he takes to hiding during the day from his creditors and so becomes a night-bound, furtive thing. The vampire bat, soon after it was first described in South America in the sixteenth century, was gleefully assimilated into preexisting vampiric folklore, ready for Bram Stoker to seize for his *Dracula* in 1897: "a tall old man, clean shaven save for a long white mustache, and clad in black from head to foot, without a single speck of colour about him anywhere." (Something people forget: Stoker's Dracula was a senior citizen.) A bat on your heraldic shield meant an awareness of the powers of chaos; it signaled your kinship with pandemonium.

But there is grace, too, in the dark: Navajo folklore holds bats as the earliest of the world's creatures, created while the whole world was in blackness, flying in the company of twelve insects through the newborn unlit earth. And for bats, of course, there is no such thing as a darkness too dark, because of one of the finest pieces of evolutionary finesse: echolocation.

It took a long time for us to learn what allows bats to move with such confidence through the night. For centuries, it was thought they merely had spectacular night vision, which led to more folk remedies; in parts of the American Midwest, it was hoped that submerging your eyes in bat blood would let you see in the dark. But bats—though they have excellent eyesight in daylight, three times sharper than our own—are not seeing through the dark. They are hearing it.

In 1793, the Italian scientist Lazzaro Spallanzani discovered that his pet owl, if he blew out his reading candle, flew straight into the wall; the local bats, though, could navigate just as swiftly and skillfully as in light. Intrigued, he covered the eyes of three wild bats with sticky birdlime, with the same result; then he removed the eyeballs of one bat, and found it "flew quickly . . . with the speed and sureness of an uninjured bat . . . My astonishment at this bat which absolutely could see although deprived of its eyes is inexpressible." In response, a Swiss zoologist, Charles Jurine, set about plugging bats' ears with wax, and found they flew like drunks; after some gruesome experiments involving paraffin and piercing the bats' inner ears, the two concluded that hearing must be vital to a bat's flight. It was, though, another hundred years before echolocation began to be truly understood.

The bat, flying through the night, sends out a screech of such high frequency that it makes our human sound waves look seashore-slow and lackadaisical. The echo, bouncing off the objects around it, returns to the bat's ears and allows it to form a picture built from the sound wave, down to details of a passing insect's hairiness and leg count. Bats are virtuosic mathematicians: in order to work out how far away a target is, they assess the time delay between sending out the call and receiving the echo, based on an innate understanding of the speed of sound. When a bat begins to hunt, its initial pulses are slower, as it scopes about for prey; but as soon as an insect is located, the calls become a blur of high-intensity ultrasonic pulses of up to 200 kilohertz: waves of two hundred thousand cycles per second. The echo produced is precise enough to avoid wires thinner than a human hair. We have no hope of ever eavesdropping on them. The average human range of hearing is a puny 20 hertz to 20 kilohertz; for bats, the range is colossal, from 11 to 212 kilohertz—but if we could hear what the bat hears, we would be deafened. The calls they send out into the world are loud: the greater bulldog bat can scream at 140 decibels, equivalent to your standing a hundred feet from a jet engine amid a rock concert.

They are such connoisseurs of darkness, it is insulting to think they would blunder into your hair. Their echolocation allows them to catch a midge the size of a pinhead: they can see you, and will avoid you. But some, of course, that are sick or dazed or moving in great swarms, have ended up tangled against the scalp, and we've caught at the idea with horror and fascination. In French folklore, a bat in a woman's hair was a harbinger of a tempestuous and terrible love affair; in Ireland, it was a sign you were on your way to hell. For those who have studied their methods, though, they can offer up marvels. Daniel Kish, a blind entrepreneur in California, taught himself to echolocate as a child, with clicks of the tongue: walking through suburban California, he is able to detect trees and walls, and ride a bicycle, navigating from a picture of the world built in pulses.

Bats look like nothing else on earth, with their mouse-faces and translucent wings. Aristotle believed that bats were birds. Then he changed his mind, concerned with their lack of a tail, and their mouselike feet: in his *De Partibus Animalium*, he put them in the middle ground—the space for those inconvenient animals who belong to two classes at the same time, along with the seals, with their fin-feet, and ostriches, which are birds with "feathers that are like hairs and useless for flight." But Aristotle did them an injustice: they are wholly, and plentifully, themselves. There are more than 1,100 species of bat in the world, making up more than a quarter of the count of all mammal species. And they are the under-sung ravishments of the night: their family contains the brown long-eared bat, with ears half as long as its body; the pied bat, which looks like a tiny winged badger; and the Honduran white bat, with a white body and yellow ears and nose. There is the painted bat in Thailand, with a body so orange it's luminous, known as the "butterfly," and the smallest bat in the world, the Kitti's hog-nosed bat, known as the bumblebee bat and smaller than your thumb. But our pleasure in miniature dollhouse things is killing them. They are one of the sixty-six bat species listed as critically endangered, endangered, or vulnerable: as tourism to see the tiny bats in Thai caves expands, and the caves are made more comfortable for visitors, their numbers are falling. It's the same old story, played out in many thousand ways: we whittle away at their numbers with our delight.

Bats very nearly burned Tokyo to the ground; perhaps they could have saved Hiroshima. During the Second World War, an American dentist, Lytle S. Adams, came up with a plan: he attempted to use Mexican free-tailed bats to carry small timed firebombs. The idea was that they would be sent out inside a bomb casing, which would open on a parachute, letting millions of bats burst out, to fly free and roost on wooden buildings in Tokyo. When the timers went off, they would set light to the whole city. Bats, Adams said, were ordained "by God to await this hour to play their part in the scheme of free human existence, and to frustrate any attempt of those who dare desecrate our way of life." Roosevelt said of him: "This man is not a nut. It sounds like a perfectly wild idea, but is worth looking into." It turned out to be more complicated than Adams expected—there was a mistake in the moment in which they were released, and instead of flocking across a wide area, they all bunched together, roosted under a fuel tank, and exploded the testing center. The plan was abandoned, in favor of a new idea: the atomic bomb.

THE

Tuna

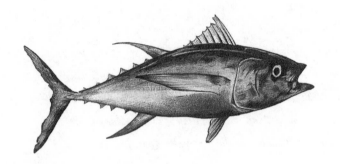

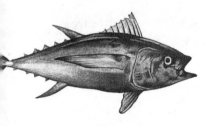

E rnest Hemingway thrilled to the tuna: to their size, and their strength. They are big as a grizzly bear, and he loved them for it. Most reach around six feet, but the largest outliers of the largest species, the Atlantic bluefin, can be twice that, and more than fourteen hundred pounds. In 1922, watching a school of tuna hunt a run of sardines from the Spanish port of Vigo, Hemingway wrote in a newspaper dispatch about a "big tuna who breaks water with a boiling crash and . . . falls back into the water with the noise of a horse diving off a dock." Their colossal heft allowed him to conceive of fishing for tuna as a heroic struggle, pure masculinity versus the ocean. He wrote:

> . . . if you land a big tuna after a six-hour fight, fight him man against fish when your muscles are nauseated with the unceasing strain, and finally bring him alongside the boat, green-blue and silver in the lazy ocean, you will be purified and be able to enter unabashed into the presence of the very elder gods, and they will make you welcome.

This is the prose of a man who longs in his deepest heart to punch fish straight out of the sea. The tuna was the fish for him: had it been possible, he would have dressed it in boxing gloves and a pair of tiny little shorts.

In Papuan mythology, the tuna is the father of the sun. In the story a woman, playing in the water with a vast tuna, felt it rub against her leg. Over time, the leg began to swell, until she cut open the swelling and from it came a baby. The child, Dudugera, "leg child," was mocked by the other children, and became aggressive and angry, a fighter; fearing for his safety, his mother took him back to the water to return him to his father. The great tuna appeared and took the boy in his mouth. But before he could be taken down into the water with his father, Dudugera told his mother to hide, because he was going to become the sun. Dudugera climbed into the sky, scorching the earth and everything on it. But to mitigate his destructive power, his mother tossed lime in the face of the sun as it rose one morning, which formed clouds and protected the world from his ferocity. The tuna has its place in stories that are large and wild, and set at the beginning.

Their name means "dart along"; they are torpedoes in the water. Of the fifteen species, the ones you are most likely to find in tins in supermarkets are the skipjack, albacore, and yellowfin, but it's the Atlantic bluefin who is the grandest, the swiftest and largest. They are midnight-blue shading to silver on top, and shining white beneath. Swimming at speed, the bluefin's top fins retract into their bodies, and they pelt at forty-three miles an hour, faster than a great white shark. So perfectly evolved are they for powering through the ocean, Pentagon-funded scientists have used the tuna body-shape as a model for the U.S. Navy's underwater missiles. They look large enough for a child to fit inside, Jonah and the Whale—

style. Atlantic bluefins swim in vast shoals of five hundred and more: to witness this, in all its speed and frothing water, is akin to seeing a migration of stampeding oceanic buffalo.

Like Hemingway's "elder gods," Atlantic bluefins do not acknowledge borders. Born in the Mediterranean Sea or the Gulf of Mexico, they grow up to hunt across the entirety of the Atlantic—shuttling from Miami to Iceland, Mauritania to Cuba, back and forth, inexhaustible. They can cross the Atlantic Ocean in just forty days, but to mate, they will commute in their great jostling eager school back to the waters near their birth spot. Exactly how they know where to go, we aren't sure: their sense of smell is remarkable, and perhaps that allows them to build an olfactory map of the ocean—or they may use the stars, or the Earth's magnetic field. We know only that, each mating season, they return for the "broadcast spawning": large groups of males and females simultaneously release eggs and sperm into the water in a hopeful cascade and leave them to fare as best they can. The vast majority of the ten million eggs a female produces a year will never be fertilized, but those that are will hatch two days later, barely the size of an eyelash. It's an unusually precarious beginning for a life that can last forty years, if we, or a very few species of sharks and toothed whales, don't catch them first.

Unlike the vast majority of fish, tuna are warm-blooded. A unique blood vessel structure allows them to store the heat they generate from movement rather than losing it to the ocean, which means that they need not depend on the water around them for their body temperature. As a result, they're as persistent as they are fast; their ability to tolerate extreme shifts in water temperature allows them to pursue their prey into the pitch-black icy depths three thousand feet down. As the water grows colder, other fish grow sluggish and hesitant; the tuna, following, can easily outstrip its groggy food. (This is not always helpful for those humans who eat tuna. The tuna's diet of hundreds of smaller fish, which they swallow whole—herring, sardines, mackerel, all of which have small amounts of mercury in their bodies—means that mercury accumulates in the tuna's flesh over its lifetime, and is never expelled. As a rule of thumb for those who want to avoid mercury poisoning: eat tinier fish, lower in the food chain.)

Our appetite for all forms of tuna goes back a very long way. We designed elaborate traps for them across Europe as early as the first century CE, making a maze of nets that catch them during their spawning period. But it was only in the aftermath of the Second World War, as our desire for the fish skyrocketed, that we became so deadly efficient at trawling for them, and so willing to destroy the ocean floor in our quest for them. Much of our fishing is on longlines—lines with baited hooks which stretch more than fifty miles along the ocean floor, catching fish indiscriminately, discarding anything unprofitable. Dolphins, who often swim alongside tuna, are collateral damage: three hundred thousand whales and dolphins are caught and discarded every year as "bycatch" of industrial fishing. The water is full of corpses. ("Dolphin safe" labels on our tins are reckoned among marine scientists to mean next to nothing: the carnage takes place miles out at sea, where regulation cannot be constant, and inspectors can be bribed.) By some estimates, ninety percent of the largest predatory fish—the megafauna, like the tuna and the shark and the swordfish with its bill long and sharp enough to kill a man—have already disappeared from the ocean. Our hunger is only growing. Kiyoshi Kimura, owner of the Sushi Zanmai sushi chain, paid $3.1 million for a 613-pound Atlantic bluefin, a world record. It's generally reckoned the bidding was artificially inflated to spark press attention and fish-based fanfare, but even so, the bluefin is one of the most valuable living things on the planet.

At the high-end Japanese restaurant chain Nobu—part-owned by Robert De Niro—with its dazzle and soft lighting in London, New York, and LA, it is possible to buy bluefin tuna. The London restaurant menu at the time of writing includes a slimy little asterisk: "Bluefin tuna is an environmentally threatened species—please ask your server for an alternative"—sashimi with a sauce of cognitive dissonance, as if that were enough to absolve the restaurant of their part in the supply chain. Trevor Corson, a former fisherman and author of the book *The Story of Sushi*, is skeptical about why we fish it at all: most people cannot tell the difference, in a blind taste, between bluefin and yellowfin. For many diners at Nobu, though, the asterisk is presumably not so much a deterrent as a victory flag: it's their scarcity that makes eating them so visceral a thrill. In the fifteenth century, Lorenzo de' Medici would from time to time turn Piazza della Signoria in Florence into a hunting field. A host of exotic animals would be collected and unleashed in the square in order to be massacred. It's a similar impulse in Nobu: devouring something rare.

Our Medici-esque edge has set them on the road to their end. The Mitsubishi conglomerate controls a forty percent share of the world market in bluefin tuna; they are freezing and hoarding huge stocks of the fish every year. While they claim this is to smooth supply on a year-to-year basis, conservationists believe they are acting in the expectation that in the event of the fish's extinction in the wild, prices will skyrocket. Frozen in great stacks at −76°F by the same company who made my childhood cassette player, the bodies would be sold for astronomical prices.

It has a name, this uniquely vile game: it is called extinction speculation. It's practiced by those who collect Norwegian shark fin, rare bear bladders, and rhino horn; men and women with hearts that sing along only to the song of money. There are collectors known to be building up huge piles of tiger pelts and vats of tiger bone wine. (The wine is made by soaking portions of a tiger's skeleton in rice wine; it takes eight years to ferment, and can then be stored indefinitely.) If tigers go extinct in the wild, which is wholly possible by 2050, the value of these assets will soar. Already, progress is looking good for those who bet on obliteration: the narrow-striped South

China tiger has not been seen in the wild since the 1980s; the Caspian tiger, which had the thickest, most luxuriant fur of all tiger subspecies, became extinct in the wild at the end of the twentieth century. A study found that, in the case of the rhinoceros, "profit-maximizing individuals may have an incentive to subsidize the slaughter of rhinos until the wild stock collapses." Poachers have been paid to shoot even those wild rhinos without marketable horns, in order to hurry along the final death.

Extinction isn't just happening because of our inertia: it's incentive-driven. The tuna migrate across the vast blue world, and up above them the gamblers watch, keeping their stocks close and secret, and waiting for the end to come.

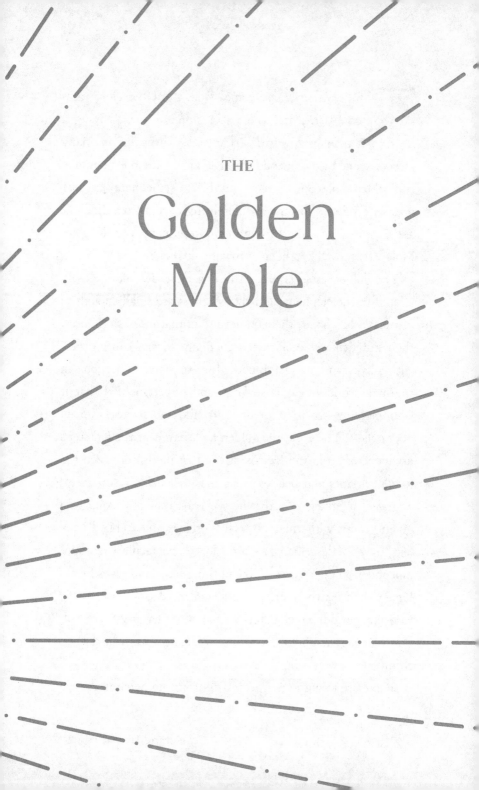

THE

Golden
Mole

The word *iridescent* comes from the Greek for "rainbow," *iris*, and the Latin suffix *-escent*, "having a tendency toward." Iridescence turns up in many insects, some birds, the odd squid—but in only one mammal, the golden mole. Some species are black, some metallic silver or tawny yellow, but under different lights and from different angles, their fur shifts through turquoise, navy, purple, gold. Moles, then, with a tendency toward sky colors.

The golden mole is not, in fact, a mole. It's more closely related to the elephant, and though most are small enough to fit in a child's hand, their bodies are miniature powerhouses: their kidneys are so efficient that many species can go their entire lives without drinking a drop of water. The bone in the mole's middle ear is so large and hypertrophied that it is immensely sensitive to underground vibrations; waiting under the soil or sand, the golden mole can hear the footsteps up above of birds and lizards; it can distinguish between the footfall of ants and beetles. There is a scientific term, *autapomorphic*, referring to a distinctive trait that is unique to a given group: with their powerful forelegs and webbed back feet, they are described by scientists as "spectacularly autapomorphic." They have been like nothing but themselves for far longer than us: there are fossil specimens dating back to the Miocene period, which extends from about twenty-three million years ago to about five million years ago. They have been shining for a very long time.

There are twenty-one species, all from sub-Saharan Africa;

as is the way of these things, many of them are named after men. There's Grant's golden mole (just three inches long, found only in the Namibian desert, known as the "dune shark") and Marley's golden mole (reddish-brown, found only on two small patches of land on the eastern slopes of the Lebombo Mountains), the robust golden mole (not robust at all: it's dying out due to sweeping habitat loss in South Africa), and the largest species, the giant golden mole. At nine inches in length, it could take all the others in a fight, but is no match for our destructive impulses, our forest clearances and mining, and is the most endangered of the golden moles. Of the twenty-one species, more than half are currently threatened with extinction due to pollution and loss of habitat: if we lose them, we will have lost the world's only rainbow mammal, a stupidity so grotesque we could not expect to be forgiven.

And then there is the twenty-first species, the Somali golden mole, the Bigfoot of golden moles, which has never been seen alive. In 1964, Alberto Simonetta, a professor at the Florence Institute of Zoology, was sifting through the contents of a disused bakery oven in Somalia. A family of barn owls had been living in it, and inside one of the pellets, Simonetta found the assorted bones of a golden mole, including the "right ramus of the lower jaw." This tiny jawbone, barely larger than a thumbnail clipping, did not match that of any other known mole, and so the *Calcochloris tytonis*—the Somali golden mole—was added to the species list. Simonetta hunted for the living version, promising children a shilling for every specimen produced, but none was forthcoming, and the only record of its life is his twenty-nine-page paper, "A New Golden Mole from Somalia with an Appendix on the Taxonomy of the Family Chrysochloridae." The Somali golden mole is listed on the International Union for Conservation of Nature's Red List of Threatened Species as "data deficient." This is a category applied to fourteen percent of all mammals: a reminder that we do not know what shares the world with us, and in what numbers. We know very little about any golden mole, but about the Somali golden mole we know nothing at all, not what color it is, nor whether it's quietly abundant in some small and as yet unsurveyed patch of land, nor if the owl ate the last one.

Perhaps the greatest mystery is why the golden mole has evolved to glow. Iridescence occurs when an object's physical structure causes light waves to combine, seeming to shift between multiple colors: the phenomenon is well represented in the natural world, but there always appears to be an obvious purpose for it. The *Morpho* butterfly, for instance, has a sweep of blue across its wings so bright and so complex we haven't yet been able to replicate it in our inks and paints; it's thought to use its iridescence to communicate with other *Morphos* over long distances, reflecting ultraviolet light with its wings. The male rufous hummingbird has an iridescent orange bib, like a Renaissance ruff rendered in Technicolor; when trying to attract a mate, it fluffs up its neck feathers and soars into the sky, then dives down so fast you can hear the air part around it. Once, drunk and on Valentine's day, I tried to share a portion of fried chicken with a club-footed London pigeon—and saw in the streetlight for the first time the way the iridescent plumage on their necks, caught at the right angle, shifts between teal and magenta.

All of these creatures are iridescent for a reason. But the golden mole is blind. Its eyes are covered with a layer of skin and fur, and it has never seen its own radiance. It lives almost entirely underground, in the cooler depths of earth and sand twenty inches below the surface, emerging only to hunt for insects. It's currently thought that the fur evolved to be densely flattened, hard-wearing, and low-friction to make burrowing easier. The iridescence is an accidental by-product. It is glory without necessary purpose, cast up by the world's slow finessing. So they burrow and breed and hunt, live and die under the African sun, unaware of their beauty, unknowingly glowing.

They are not alone in this. The human being, too, is a shining thing. We are infinitesimally bioluminescent—chemical reactions within the human body cast up photons, the elementary particle of light. The light emitted is a thousand times weaker than human eyesight, but it is constant, and clustered around the face. Like the golden mole, we too have a radiance invisible to ourselves.

THE

Human

The greatest lie that humans ever told is that the Earth is ours, and at our disposal. It's a lie with the power to destroy us all.

We must cease from telling that lie, because the world is so rare, and so wildly fine. It is populated with such strangenesses and imperiled astonishments.

Among them, human attention is perhaps one of the rarest and finest. So this book has been a wooing. It has been an asking for your wonder, and for your attention. Fear and rage will help galvanize us, but they will not suffice alone: our competent and furious love will have to be what fuels us. For what is the finest treasure? Life. It is everything that lives, and the earth upon which they depend: narwhal, spider, pangolin, swift, faulted and shining human. It calls out for our far more urgent, more iron-willed treasuring.

• • •

To end, a story.

It is a story about humans and our reckonings with treasure; the story of the Sibylline books. The books were a collection of oracular sayings, written in Greek poetry in around 510 BCE. The tradition runs that an elderly sibyl—a prophetess—offered the last king of Rome the chance to buy nine books containing the prophecies of the world. The apparently true story of their purchase has been retold many hundreds of times, most famously by the Roman grammarian Aulus Gellius in his *Noctes Atticae* (*Attic Nights*), c. 177 CE; by Origen of Alexandria the following century; and by the late, great Douglas Adams, in his *Last Chance to See*, on whose telling this is based. The story loosely (very loosely) runs like this:

There once was a great and flourishing city, with feasts and hard work and citizens leading busy lives. One spring an old woman came to the city, dressed in hard-worn cloth and strong shoes. She had with her nine books, which contained all the wisdom and knowledge, all the as-yet-untold secrets of the world. She would, she said, sell the nine books, for the price of one large sack of gold. (Aulus Gellius doesn't specify the amount: but it is "an immense and exorbitant sum.")

The city's people found this both mildly hilarious and amorphously annoying; this woman, they said, had very little sense of economics or value or gold itself, and they suggested she take her books and go.

"As you wish," she said—but first, she would burn three of the books.

She built a small neat fire in the square, burned three of the volumes in which all the secrets of the world were contained, and went on her way with the smoke still in the air.

It was a bad winter that year, with flooding and snowstorms, but the citizens of the city still thrived, more or less. As the summer began to shine again, the old wise woman returned.

How, the people inquired, was the undiscovered-secrets-of-the-world peddling going?

Fine, said the old woman: and she would sell them the six remaining books: two-thirds of all the world's wisdom and secrets. The price, though, was higher: two sacks of gold.

This, the citizens pointed out, was outrageous profiteering: she couldn't double the price for two-thirds of the wisdom. The old woman shrugged, and asked to borrow a match. Three more books went up in flames.

Another winter came, and this one was harder, and colder, and there were more deaths than the people could easily bear, but when the sun came things were better and the citizens were encouraged to put it out of mind.

The old woman arrived, with three books in her bag. They could have them, she said, for four sacks of gold. The people, whose mathematical skills were impeccable, laughed uneasily. She couldn't be serious?

The old woman asked for firewood.

"Wait!" said the city's people: perhaps it would be worth having a look, at least. Leave the books with them: they could have a series of debates and consultations, and, at some non-specific point in the future, they would let her know if they'd agreed whether or not there was anything worthwhile in it.

The old woman shook her head. "If I could trouble you for that firewood."

They refused to bring the wood.

"You don't want the books?"

"Not at that price: we can't afford it. You need to be realistic."

So the old woman, who was realistic to her very core, shrugged. She gathered a heap of dried grass left over from the haymaking—which had been poor that year—thrust two books into the center, and set them alight. They burned fast.

When she returned the following spring, the single book under her arm, the city's people were waiting for her. "We know," they said. "Eight sacks of gold. We've got it here."

"The price," said the old woman, "is sixteen sacks of gold."

"But," said the populace, "we've planned and budgeted for eight!"

"Sixteen is cheap," she said. "You're being ridiculous!"

The old woman looked at them with the full force of her eyebrows, and the wiser among the people shrank back. "It is cheap. The book contains gold beyond all gold."

"It's been a hard year! We're struggling."

The old woman, who was collecting kindling at a surprisingly swift pace, said nothing.

The citizens ran back to their homes, and argued furiously, and in the end they gathered the gold. They dragged sixteen sacks back to the woman's pile of sticks, which she had just topped with the last remaining book.

They seized it, with hunger and hope and desperation.

The old woman nodded, and loaded the gold sacks onto two strong horses, and turned to leave the city.

"It had better," they called to her retreating back, "be worth the price."

"Of course it is," said the old woman. "Of course. It's astonishing. It's gold beyond all gold."

She reached the gates of the city. She spoke without turning around. "You should have seen what was burned."

● ● ●

So much can still be saved. It is the greatest task, now, of everyone alive: to keep it from the flames.

Author's Note

When this book first came out in the UK, I was often asked, in response: What, then, can I do? Every time I was asked, I thought of John Maynard Keynes, the father of macroeconomics and proud owner of a mustache like a shoe brush.

In 1930, Keynes wrote a prediction for the future. By 2030, he prophesied in his essay "Economic Possibilities for Our Grandchildren" that technological advance will have largely displaced human labor. Our standard of living would be so high as to free us to discover, for the very first time in human history, ways to live well. Then, "the love of money as a possession . . . will be recognized for what it is, a somewhat disgusting morbidity." It would be seen as a "semi-criminal, semi-pathological propensity." It is only, he said, when the accumulation of wealth is no longer the central impulse of humanity, that "we shall be able to rid ourselves of many of the pseudo-moral principles which have hag-ridden us for two hundred years, by which we have exalted some of the most distasteful of human qualities into the position of the highest virtues."

No oracle, Keynes: too optimistic about us. But he was utterly right in this: that a dramatic recalibration will be needed. We will need the decoupling of desire and conspicuous consumption, a new way of thinking about power, community, joy, luxury. History doesn't suggest it will be easy, but it does tell us that it is possible. We humans have shown ourselves capable of change so bold it could knock the breath out of you. Why should we follow the old ways? Were they so perfect as to leave no room for something bolder, tougher, wittier, more equal, and more just?

Speaking with scientists and activists over several years, this is how I've been told we can act:

Vote. Engage actively in politics at every level, and clamor for legislation. Vote for political parties that believe without caveat in the existence of manmade climate change and in taking action to move immediately away from fossil fuels and toward wind and solar power. Call for laws that will make fossil fuel companies and polluting nations pay for the damage they have done.

If you are on the side of the equation that has money to invest, the most urgent thing you can do is make sure that no vestige of it is invested in fossil fuels.

Educate yourself. Educate others. Refuse to stop talking about it. The world deserves your clear-sighted and nuanced awe. The only possible way not to lose sight of the true miracle of the world, and of its vulnerability, is by learning: by building knowledge upon knowledge.

Climate change must be fought at the global, political, governmental level. But if you, like me, belong to the part of the world that consumes most at the expense of the destruction of those parts of the world that consume least, there are personal actions that are worth taking. Eat less or no meat, buy vastly less, throw away less, treat domestic flights as the behavior of the malarially unhinged. Which is to say: we can participate in the process of reimagining. Invent new ways to relish the world that do not involve the relentless purchasing of new things.

Protest. Protest both when you believe it will work—for we have vivid evidence of when it has worked—and when you do not. The farmer-poet-activist Wendell Berry, in his book *What Are People For?*, wrote: "Protest that endures, I think, is moved by a hope far more modest than that of public success: namely, the hope of preserving qualities in one's own heart and spirit that would be destroyed by acquiescence."

And finally, avoid the temptation to strike a pose and say "It's too late." It's not true. It's half-baked nihilism: and like all half-baked nihilism, it lazily tempts us to think that what we do doesn't matter now. It does matter: it will always matter. The time to give up is never. Hope—active, purposeful, informed hope—is what we owe the world. A body of unimaginable splendor turns on its axis, calling us to its aid.

About This Book

Half of all the author royalties from this book will go in perpetuity to charities working to push back at climate change and environmental destruction, one land-based, one sea-based. In buying it, you support them, for which I thank you most truly.

Acknowledgments

My greatest thanks are owed to Mary-Kay Wilmers, editor and co-founder of the *London Review of Books,* and to Alice Spawls, her successor as co-editor. Some of these animals first appeared there, and I am so grateful to the paper for giving them a home.

I owe enormous thanks, too, to Vikrom Mathur, for his generosity and scientific expertise, and to Amy Jeffs, who long ago told me about the existence of the golden mole. And there are so many others, to whom I owe such gratitude and love that they would fill another book. To those, who know who they are: thank you.

Further Reading

The Wombat

John Simons, *Rossetti's Wombat: Pre-Raphaelites and Australian Animals in Victorian London* (2008).

Nicole Starbuck, *Baudin, Napoleon and the Exploration of Australia* (2015).

Larry Vogelnest and Rupert Woods, eds., *Medicine of Australian Mammals* (2008).

The Greenland Shark

J. D. Borucinska et al., "Ocular Lesions Associated with Attachment of the Parasitic Copepod Ommatokoita Elongata (Grant) to Corneas of Greenland Sharks, Somniosus Microcephalus," *Journal of Fish Diseases* 21, no. 6 (1998): 415–22.

Julius Nielsen, Jan Heinemeier et al., "Eye Lens Radio-Carbon Reveals Centuries of Longevity in the Greenland Shark (Somniosus Microcephalus)," *Science* 353 (2016): 702–4.

Morten Strøksnes, *Shark Drunk: The Art of Catching a Large Shark from a Tiny Rubber Dinghy in a Big Ocean* (2017).

The Raccoon

Matthew Costello, "Raccoons at the White House," *White House Historical Association* (June 8, 2018).

Daniel Heath Justice, *Raccoon* (2021).

L. Stanton, E. Davis, S. Johnson et al. "Adaptation of the Aesop's Fable Paradigm for Use with Raccoons (*Procyon lotor*): Considerations for Future Application in Non-Avian and Non-Primate Species," *Animal Cognition* 20 (2017): 1147–52.

The Giraffe

Michael Allin, *Zarafa: The True Story of a Giraffe's Journey from the Plains of Africa to the Heart of Post-Napoleonic France* (1999).

Matilda E. Dunn et al., "Investigating the International and Pan-African Trade in Giraffe Parts and Derivatives," *Conservation Science and Practice* 3, no. 5 (2021): e390.

Bryan Shorrocks, *The Giraffe: Biology, Ecology, Evolution and Behaviour* (2016).

The Swift

Susanne Åkesson et al., "Migration Routes and Strategies in a Highly Aerial Migrant, the Common Swift *Apus apus*, Revealed by Light-Level Geolocators," *PLoS ONE* 7, no. 7 (2012): e41195.

Anders Hedenström et al., "Annual 10-Month Aerial Life Phase in the Common Swift Apus apus," *Current Biology* 26, no. 22 (2016): 3066–70.

David and Andrew J. Lack, *Swifts in a Tower* (2018).

The Lemur

Peter M. Kappeler and J. Ganzhorn, *Lemur Social Systems and Their Ecological Basis* (2013).

Ivan Norscia and Elisabetta Palagi, *The Missing Lemur Link: An Ancestral Step in the Evolution of Human Behaviour* (2016).

Elwyn Simons and David M. Meyers, "Folklore and Beliefs About the Aye Aye (Daubentonia madagascariensis)," *Lemur News* 6 (2001): 11–16.

The Hermit Crab

Jennifer E. Angel, "Effects of Shell Fit on the Biology of the Hermit Crab *Pagurus longicarpus* (Say)," *Journal of Experimental Marine Biology and Ecology* 243, no. 2 (2000): 169–84.

Aleksandr Mironenko, "A Hermit Crab Preserved Inside an Ammonite Shell from the Upper Jurassic of Central Russia: Implications to Ammonoid Palaeoecology," *Palaeogeography, Palaeoclimatology, Palaeoecology* 537 (2020): e109397.

Judith S. Weis, *Walking Sideways: The Remarkable World of Crabs* (2012).

The Seal

Gisela Heckel and Yolanda Schramm, eds., *Ecology and Conservation of Pinnipeds in Latin America* (2021).

Colleen Reichmuth and Caroline Casey, "Vocal Learning in Seals, Sea Lions, and Walruses," *Current Opinion in Neurobiology* 28 (2014): 66–71.

Marianne Riedman, *The Pinnipeds: Seals, Sea Lions, and Walruses* (1990).

The Bear

Thomas Browne, *Pseudodoxia Epidemica or Enquiries into very many received tenents and commonly presumed truths* (1646).

Caroline Grigson, *Menagerie: The History of Exotic Animals in England* (2015).

Terence Hawkes, *Shakespeare in the Present* (2002).

The Narwhal

Zackary Graham et al., "The Longer the Better: Evidence That Narwhal Tusks Are Sexually Selected," *Biology Letters* 16, no. 3 (2020).

Geir Hønneland and Leif Christian Jensen, *Handbook of the Politics of the Arctic* (2015).

Martin Nweeia et al., "Sensory Ability in the Narwhal Tooth Organ System," *The Anatomical Record* 297, no. 4 (2014): 599–617.

The Crow

Heather Cornell et al., "Social Learning Spreads Knowledge About Dangerous Humans Among American Crows," *Proceedings of the Royal Society* 279 (2011): 499–508.

Nathan Emery and Frans B. M. Waal, *Bird Brain: An Exploration of Avian Intelligence* (2016).

Mark Walters, *Seeking the Sacred Raven: Politics and Extinction on a Hawaiian Island* (2012).

The Hare

P. J. Edwards et al., "Review of the Factors Affecting the Decline of the European Brown Hare, *Lepus europaeus* (Pallas, 1778) and the Use of Wildlife Incident Data to Evaluate the Significance of Paraquat," *Agriculture, Ecosystems and Environment* 79, no. 2–3 (2000): 95–103.

Alan Kors and Edward Peters, eds., *Witchcraft in Europe, 400–1700* (2001).

Marianne Taylor, *The Way of the Hare* (2017).

The Wolf

Barry Lopez, *Of Wolves and Men* (revised edition, 2004).

L. David Mech and Luigi Boitani, eds., *Wolves: Behavior, Ecology, and Conservation* (2003).

Alanna Skuse, "Wombs, Worms and Wolves: Constructing Cancer in Early Modern England," *Social History of Medicine* 27, no. 4 (2014): 632–48.

The Hedgehog

Peter Brears, *Cooking and Dining in Medieval England* (2008).

Elizabeth Morrison, ed., *Book of Beasts: The Bestiary in the Medieval World* (2019).

Pliny the Elder, *Natural History*, trans. H. Rackham, 5 vols. (2012).

The Elephant

Prithiviraj Fernando et al., "DNA Analysis Indicates That Asian Elephants Are Native to Borneo and Are Therefore a High Priority for Conservation," *PLoS Biology* 1, no. 1 (2003): e6.

Michael Garstang, *Elephant Sense and Sensibility* (2015).

Henry Du Pré Labouchère, *Diary of the Besieged Resident in Paris* (1871).

The Seahorse

David Abulafia, *The Boundless Sea: A Human History of the Oceans* (2020).

T. R. Consi et al., "The Dorsal Fin Engine of the Seahorse (Hippocampus sp.)," *Journal of Morphology* 248, no. 1 (2001): 80–97.

Anthony B. Wilson et al., "The Dynamics of Male Brooding, Mating Patterns, and Sex Roles in Pipefishes and Seahorses (family Syngnathidae)," *Evolution* 57, no. 6 (2003): 1374–86.

The Pangolin

Daniel Ingram et al., "Assessing Africa-wide Pangolin Exploitation by Scaling Local Data," *Conservation Letters* 11, no. 2 (2018): e12389.

Bin Wang et al., "Pangolin Armor: Overlapping, Structure, and Mechanical Properties of the Keratinous Scales," *Acta Biomaterialia* 41 (2016): 60–74.

Carly Waterman et al., eds., *Pangolins: Science, Society and Conservation* (2019).

The Stork

Tim Birkhead et al., *Ten Thousand Birds: Ornithology Since Darwin* (2014).

Thomas Harrison, "Birds in the Moon," *Isis* 45, no. 4 (1954): 323–30.

Otto Lilienthal, *Birdflight as the Basis of Aviation: A Contribution Towards a System of Aviation, Compiled from the Results of Numerous Experiments Made by O. and G. Lilienthal* (1889).

Isabella Tree, *Wilding: The Return of Nature to a British Farm* (2018).

The Spider

Leslie Brunetta and Catherine L. Craig, *Spider Silk: Evolution and 400 Million Years of Spinning, Waiting, Snagging, and Mating* (2010).

Norman Platnick, *Spiders of the World: A Natural History* (2020).

Raimondo Maria de Termeyer, revised by Burt Green Wilder, *Researches and Experiments upon Silk from Spiders, and upon Their Reproduction* (1866).

The Bat

Arden Christen and Joan Christen, "Dr. Lytle Adams' Incendiary 'Bat Bomb' of World War II," *Journal of the History of Dentistry* 52, no. 3 (2004): 109–15.

M. Brock Fenton and Nancy B. Simmons, *Bats: A World of Science and Mystery* (2015).

George D. Pollak and John H. Casseday, *The Neural Basis of Echolocation in Bats* (2012).

The Tuna

Trevor Corson, *The Story of Sushi: An Unlikely Saga of Raw Fish and Rice* (2008).

Charles F. Mason, Erwin H. Bulte, and Richard D. Horan, "Banking on Extinction: Endangered Species and Speculation," *Oxford Review of Economic Policy* 28, no. 1 (2012): 180–92.

Jennifer Telesca, *Red Gold: The Managed Extinction of the Giant Bluefin Tuna* (2020).

The Golden Mole

Richard Girling, *The Hunt for the Golden Mole: All Creatures Great and Small, and Why They Matter* (2014).

M. J. Mason and P. M. Narins, "Seismic Sensitivity in the Desert Golden Mole (Eremitalpa granti): A Review," *Journal of Comparative Psychology* 116, no. 2 (2002): 158–63.

J. D. Skinner and Christian T. Chimimba, *The Mammals of the Southern African Sub-region* (2005).

The Human

Douglas Adams and Mark Carwardine, *Last Chance to See* (1989).

Wendell Berry, *What Are People For?* (2010).

Frantz Fanon, *The Wretched of the Earth* (1961).

Aulus Gellius, *The Attic Nights, with an English Translation*, trans. and ed. John C. Rolfe (1927).

Amitav Ghosh, *The Great Derangement: Climate Change and the Unthinkable* (2016).

Alexis Pauline Gumbs, *Undrowned: Black Feminist Lessons from Marine Mammals* (2020).

bell hooks, *All About Love: New Visions* (2000).

Steve Rayner and Elizabeth L. Malone, eds., *Human Choice and Climate Change*, 4 vols. (1998).

Marilynne Robinson, *The Death of Adam: Essays on Modern Thought* (2000).

Andrew E. Snyder-Beattie et al., "The Timing of Evolutionary Transitions Suggests Intelligent Life Is Rare," *Astrobiology* 21, no. 3 (2021): 265–78.

Rebecca Solnit, *Hope in the Dark: Untold Histories, Wild Possibilities* (2016).

David Wallace-Wells, *The Uninhabitable Earth: A Story of the Future* (2019).

ABOUT THE AUTHOR

KATHERINE RUNDELL is a Fellow of St. Catherine's College, Oxford. She is the author of *Super-Infinite: The Transformations of John Donne,* which won the Baillie Gifford Prize for Non-Fiction, and *Why You Should Read Children's Books, Even Though You Are So Old and Wise.* Her multi-award-winning novels for children have been translated into more than forty languages and have sold more than two million copies worldwide. She has written for, among others, *The New Yorker, The New York Times, The New York Review of Books,* the *London Review of Books,* and *The Times Literary Supplement,* about books, the natural world, and night climbing.